Lecture Notes in Computer Science 8845

Commenced Publication in 1973
Founding and Former Series Editors:
Gerhard Goos, Juris Hartmanis, and Jan van Leeuwen

Editorial Board

David Hutchison
  Lancaster University, Lancaster, UK
Takeo Kanade
  Carnegie Mellon University, Pittsburgh, PA, USA
Josef Kittler
  University of Surrey, Guildford, UK
Jon M. Kleinberg
  Cornell University, Ithaca, NY, USA
Friedemann Mattern
  ETH Zurich, Zürich, Switzerland
John C. Mitchell
  Stanford University, Stanford, CA, USA
Moni Naor
  Weizmann Institute of Science, Rehovot, Israel
C. Pandu Rangan
  Indian Institute of Technology, Madras, India
Bernhard Steffen
  TU Dortmund University, Dortmund, Germany
Demetri Terzopoulos
  University of California, Los Angeles, CA, USA
Doug Tygar
  University of California, Berkeley, CA, USA
Gerhard Weikum
  Max Planck Institute for Informatics, Saarbruecken, Germany

More information about this series at http://www.springer.com/series/7407

Jin Akiyama · Hiro Ito
Toshinori Sakai (Eds.)

Discrete and Computational Geometry and Graphs

16th Japanese Conference, JCDCGG 2013
Tokyo, Japan, September 17–19, 2013
Revised Selected Papers

 Springer

Editors
Jin Akiyama
Tokyo University of Science
Tokyo
Japan

Toshinori Sakai
Tokai University
Tokyo
Japan

Hiro Ito
The University of Electro-Communications
Tokyo
Japan

ISSN 0302-9743 ISSN 1611-3349 (electronic)
Lecture Notes in Computer Science
ISBN 978-3-319-13286-0 ISBN 978-3-319-13287-7 (eBook)
DOI 10.1007/978-3-319-13287-7

Library of Congress Control Number: 2014956235

LNCS Sublibrary: SL1 – Theoretical Computer Science and General Issues

Springer Cham Heidelberg New York Dordrecht London

Printed on acid-free paper

Springer International Publishing AG Switzerland is part of Springer Science+Business Media
(www.springer.com)

Preface

This volume consists of the peer-reviewed papers of the 16th Japan Conference on Discrete and Computational Geometry and Graphs (JCDCG^2 2013), which were held during September 17–19, 2013 at Tokyo University of Science, Tokyo, Japan.

The previous conferences were held in Tokyo as JCDCG 1997, 1998, 1999, 2000, 2002, 2004, 2006, and 2011 (Japan Conference on Discrete and Computational Geometry; including domestic conferences); in Kyoto as KyotoCGGT 2007 (Kyoto International Conference on Computational Geometry and Graph Theory); and in Kanazawa as JCCGG 2009 (Japan Conference on Computational Geometry and Graphs). Other conferences in this series were also held in the Philippines (2001), Indonesia (2003), China (2005, 2010), and Thailand (2012). The proceedings of these conferences were published by Springer as a part of the LNCS series in volumes 1763, 2098, 2866, 3330, 3742, 4381, 4535, 7033, and 8296. The proceedings of the fifth and twelfth conferences (2001 and 2009) were published by Springer-Verlag as special issues of the journal Graphs and Combinatorics, Vol. 18, No. 4, 2002 and Vol. 27, No. 3, 2011.

The organizers of JCDCG^2 2013 thank the following invited speakers, Sergey Bereg, Erik Demaine, Ferran Hurtado, Ken-ichi Kawarabayashi, David Kirkpatrick, Stefan Langerman, Janos Pach, Kokichi Sugihara, and Jorge Urrutia, for their fruitful talks. They also gratefully acknowledge the support of the conference secretariat, the speakers, and all of the participants of the conference.

September 2014

Jin Akiyama
Hiro Ito
Toshinori Sakai

Preface

Contents

Covering Partial Cubes with Zones

Jean Cardinal[1]($\boxtimes$) and Stefan Felsner[2]($\boxtimes$)

[1] Université Libre de Bruxelles (ULB), Brussels, Belgium
jcardin@ulb.ac.be
[2] Technische Universität Berlin, Berlin, Germany
felsner@math.tu-berlin.de

Abstract. A partial cube is a graph having an isometric embedding in a hypercube. Partial cubes are characterized by a natural equivalence relation on the edges, whose classes are called *zones*. The number of zones determines the minimal dimension of a hypercube in which the graph can be embedded. We consider the problem of covering the vertices of a partial cube with the minimum number of zones. The problem admits several special cases, among which are the problem of covering the cells of a line arrangement with a minimum number of lines, and the problem of finding a minimum-size fibre in a bipartite poset. For several such special cases, we give upper and lower bounds on the minimum size of a covering by zones. We also consider the computational complexity of those problems, and establish some hardness results.

1 Introduction

As an introduction and motivation to the problems we consider, let us look at two puzzles.

Hitting a consecutive pair. Given a set of n elements, how many pairs of them must be chosen so that every permutation of the n elements has two consecutive elements forming a chosen pair?

Guarding cells of a line arrangement. Given an arrangement of n straight lines in the plane, how many lines must be chosen so that every cell of the arrangement is bounded by at least one of the chosen line?

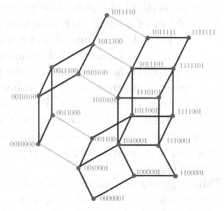

Fig. 1. A partial cube with vertex labels representing an isometric embedding in Q_7 and a highlighted zone.

While different, the two problems can be cast as special cases of a general problem involving *partial cubes*. The n-dimensional hypercube Q_n is the graph with the set $\{0,1\}^n$ of binary words of length n as vertex set, and an

© Springer International Publishing Switzerland 2014
J. Akiyama et al. (Eds.): JCDCGG 2013, LNCS 8845, pp. 1–13, 2014.
DOI: 10.1007/978-3-319-13287-7_1

edge between every pair of vertices that differ on exactly one bit. A *partial cube* G of dimension n is a subgraph of the n-dimensional hypercube with the property that the distance between two vertices in G is equal to their Hamming distance, i.e., their distance in Q_n. In general, a graph G is said to have an *isometric embedding* in another graph H whenever G is a subgraph of H and the distance between any two vertices in G is equal to their distance in H. Hence partial cubes are the graphs admitting an isometric embedding in Q_n, for some n. The edges of a partial cube can be partitioned into at most n equivalence classes called *zones*, each corresponding to one of the n directions of the edges of Q_n (Fig. 1).

We consider the following problem:

Partial Cube Covering: *Given a partial cube, find a smallest subset S of its zones such that every vertex is incident to an edge of one of the zones in S.*

If we refer to the labeling of the vertices by words in $\{0,1\}^n$, the problem amounts to finding a smallest subset I of $[n]$ such that for every vertex v of the input partial cube, there is at least one $i \in I$ such that flipping bit i of the word of v yields another vertex.

2   Classes of Partial Cubes

The reader is referred to the books of Ovchinnikov [16] and Hammack et al. [14] for known results on partial cubes. Let us also mention that some other structures previously defined in the literature are essentially equivalent to partial cubes. Among them are *well-graded families of sets* defined by Doignon and Falmagne [7], and *Media*, defined by Eppstein, Falmagne, and Ovchinnikov [10].

We are interested in giving bounds on the minimum number of zones required to cover the vertices of a partial cube, but also in the computational complexity of the problem of finding such a minimum cover. Regarding bounds, it would have been nice to have a general nontrivial result holding for every partial cube. Unfortunately, only trivial bounds hold in general.

In one extreme case, the partial cube $G = (V, E)$ is a star, consisting of one central vertex connected to $|V| - 1$ other vertices of degree one. This is indeed a partial cube in dimension $n = |V| - 1$, every zone of which consists of a single edge. Since there are n vertices of degree one, all n zones must be chosen to cover all vertices. In the other extreme case, the partial cube is such that there exists a single zone covering all vertices. This lower bound is attained by the hypercube Q_n.

Table 1 gives a summary of our results for the various families of partial cubes that we considered. For each family, we consider upper and lower bounds and complexity results. We now briefly describe the various families we studied. The proofs of the new results are in the following sections.

2.1   Hyperplane Arrangements

The dual graph of a simple[1] arrangement of n hyperplanes in $\mathbb{R}^d$ is a partial cube of dimension n. The partial cube covering problem becomes the following: given a simple hyperplane arrangement, find a smallest subset S of the hyperplanes such that every cell of the arrangement is bounded by at least one hyperplane in S.

The case of line arrangements was first considered in [3]. In fact it was this problem that motivated us to investigate the generalization to partial cubes. The best known lower and upper bounds for line arrangements on $|S|$ are of order $n/6$ and $n - O(\sqrt{n \log n})$ (see [1]). The complexity status of the optimization problem is unknown (Fig. 2).

Fig. 2. An arrangement of 4 lines superimposed with its dual.

Instead of lines and hyperplanes it is also possible to consider Euclidean or spherical arrangements of pseudo-lines and pseudo-hyperplanes, their duals are still partial cubes. Actually, all the cited results apply to the case of arrangements of pseudo-lines.

Spherical arrangements of pseudo-hyperplanes are equivalent to oriented matroids, this is the Topological Representation Theorem of Folkman and Lawrence. The pseudo-hyperplanes correspond to the elements of the oriented matroid and the cells of the arrangement correspond to the topes of the oriented matroid. Hence our covering problem asks for a minimum size set C of elements such that for every tope T there is an element $c \in C$ such that $T \oplus c$ is another tope. For more on oriented matroids we refer to [2].

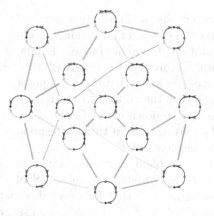

Fig. 3. Flip graph of acyclic orientations of the 4-cycle.

2.2   Acyclic Orientations

From a graph $G = (V, E)$, we can define a partial cube H in which every vertex is an acyclic orientation of G, and two orientations are adjacent whenever they differ by a single edge reversal (*flip*). This partial cube has dimension equal to the number of edges of G. It is also the dual graph of the arrangement of the $|E|$ hyperplanes of equation $x_i = x_j$ for $ij \in E$ in $\mathbb{R}^V$. Every cell of this arrangement

---

[1] An arrangement is called *simple* if any $d + 1$ hyperplanes have empty intersection.

corresponds to an acyclic orientation, and adjacent cells are exactly those that differ by a single edge. This observation, in particular, shows that H is connected. The graph H and the notion of edge flippability have been studied by Fukuda et al. [11] and more recently by Cordovil and Forge [6].

The partial cube covering problem now becomes the following: given a graph $G = (V, E)$, find a smallest subset $S \subseteq E$ such that for every acyclic orientation of G, there exists $e \in S$ such that flipping the orientation of e does not create a cycle (Fig. 3).

2.3 Median Graphs

A median graph is an undirected graph in which every three vertices x, y, and z have a unique *median*, i.e., a vertex $\mu(x, y, z)$ that belongs to shortest paths between each pair of x, y, and z. Median graphs form structured subclass of partial cubes that has been studied extensively, see [5] and references therein. Since the star and the hypercube are median graph there are no non-trivial upper or lower bounds for the zone cover problem on this class. We use a construction of median graphs to prove hardness of the zone cover problem.

2.4 Distributive Lattices

Cover graphs of distributive lattices[2] are partial cubes. From Birkhoff's representation theorem (a.k.a. Fundamental Theorem of Finite Distributive Lattices) we know that there is a poset P such that the vertices of the partial cube are the downsets of P, the zones of the partial cube in turn correspond to the elements of P.

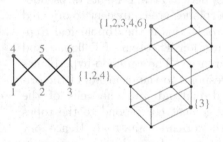

Fig. 4. A poset (left) and the cover graph of its lattice of down-sets (right).

The problem becomes: given a poset P, find a smallest subset S of its elements such that for every downset D of P, there exists $v \in S$ such that either $D \cup \{v\}$ or $D \setminus \{v\}$ is a downset, distinct from D.

2.5 Trees

Trees with n edges are partial cubes of dimension exactly n. Since every zone contains exactly one edge, the partial cube covering problem on trees boils down to the edge cover problem on trees. There are instances attaining the lower and upper bounds of $n/2$ and $n-1$, moreover, there is a simple dynamic programming algorithm that computes an optimal cover in linear time.

---

[2] For a very good introduction to terminology related to partial orders and lattices we refer to Chapter 3 of Stanley, *Enumerative Combinatorics Vol. I* [17].

Table 1. Worst-case bounds and complexities for the special cases of the partial cube covering problem. When used, n denotes the dimension of the partial cube.

Partial cube	Lower bound	Upper bound	Complexity
Arrangements of n lines in $\mathbb{R}^2$	$n - o(n)$ (Theorem 1)	$n - \Omega(\sqrt{n \log n})$ [1]	–
Arrangements of n hyperplanes in $\mathbb{R}^d$	–	$n - \Omega(n^{1/d})$ (Theorem 2)	–
Acyclic orientations	–	Minimum edge cut (Theorem 4)	Recognition is coNP-complete, even for complete graphs (Theorem 3)
Median graphs	1	n	NP-complete (Corollary 2), APX-hard (Corollary 3)
Distributive lattices with representative poset of n elements	–	$2n/3$ (Corollary 4)	Recognition is coNP-complete (Corollary 5) Σ_2^P-complete (Corollary 6)
Trees with n edges	$n/2$	$n - 1$	P (min edge cover)

3 Line Arrangements

Recall that we have defined a guarding set for an arrangement of lines as a subset of the lines n so that every cell of the arrangement is bounded by at least one of the chosen lines. We first give a lower bound on the size of a guarding set for an arrangement of lines.

Theorem 1. *The minimum number of lines needed to guard the cells of any arrangement of n lines is $n - o(n)$.*

Proof. The proof is a direct consequence of known results on the following problem from Erdős: given a set of points in the plane with no four points on a line, find the largest subset in general position, that is, with no three points on a line. Let $\alpha(n)$ be the minimum size of such a set over all arrangements of n points. Füredi observed [12] that $\alpha(n) = o(n)$ follows from the density version of the Hales-Jewett theorem [13]. But this directly proves that we need at least $n - o(n)$ lines to guard all cells of an arrangement. The reduction is the one observed by Ackerman et al. [1]: consider the line arrangement that is dual to the point set, and slightly perturb it so that each triple of concurrent lines forms a cell of size three. Now the complement of any guarding set is in general position in the primal point set. □

We now show that for arrangements of hyperplanes in $\mathbb{R}^d$, with $d = O(1)$, there always exists a guarding set of size at most $n - \Omega(n^{1/d})$. The proof is along the lines of the proof given in [3] for the $d = 2$ case.

Theorem 2. *In every arrangement of hyperplanes in general position in* $\mathbb{R}^d$ *(with* $d = O(1)$*), there exists a subset of the hyperplanes of size at most* $n - \Omega(n^{1/d})$ *such that every cell is bounded by at least one hyperplane in the subset.*

Proof. We will prove that every such arrangement has an *independent set* of size $\Omega(n^{1/d})$, where an independent set is defined as the complement of a guarding set, that is, a subset of the hyperplanes such that no cell is bounded by hyperplanes of the subset only.

Let H be a set of n hyperplanes, and consider an arbitrary, inclusionwise maximal independent set I. For each hyperplane $h \in H \setminus I$, there must be a cell c_h of the arrangement that is bounded by a set of hyperplanes $C \cup \{h\}$ with $C \subseteq I$, since otherwise we could add h to I, and I would not be maximal. If c_h has size at least $d + 1$, i.e., if c_h is incident to at least $d + 1$ hyperplanes, then there must be a vertex of the arrangement induced by I that is also a vertex of c_h. We charge h to this vertex. Note that each vertex can only be charged $2^d = O(1)$ times. If c_h has size d, we charge h to the remaining $(d - 1)$-tuple of hyperplanes of I bounding c_h. This tuple can also be charged at most $O(1)$ times. Therefore

$$|H \setminus I| = n - |I| \leq 2^d \cdot O(|I|^d) + O(|I|^{d-1})$$
$$|I| = \Omega(n^{1/d}). \qquad \square$$

4 Acyclic Orientations

Given a graph $G = (V, E)$ we wish to find a subset $S \subseteq E$ such that for every acyclic orientation of G, there exists a *flippable* edge $e \in S$, that is, an edge $e \in S$ such that changing the orientation of e does not create any cycle. Let us call such a set a *guarding set* for G. Note that an oriented edge $e = uv$ in an acyclic orientation is flippable if and only if it is not *transitive*, that is, if and only if uv is the only oriented path from u to v.

4.1 Complexity

Theorem 3. *Given a graph* $G = (V, E)$ *and a subset* $S \subseteq E$ *of its edges, the problem of deciding whether* S *is a guarding set is coNP-complete, even if* G *is a complete graph.*

Proof. The set S is not a guarding set if and only if there exists an acyclic orientation of G in which all edges $e \in S$ are transitive.

Consider a simple graph H on the vertex set V, and define G as the complete graph on V, and S as the set of non-edges of H. We claim that S is a guarding set for G if and only if H does not have a Hamilton path. Since deciding the existence of a Hamilton path is NP-complete, this proves the result.

To prove the claim, first suppose that H has a Hamilton path, and consider the acyclic orientation of G that corresponds to the order of the vertices in the path. Then by definition, no edge of S is in the path, hence all of them are transitive, and S is not a guarding set. Conversely, suppose that S is not guarding. Then there exists an acyclic orientation of G in which all edges of S are transitive. The corresponding ordering of the vertices in V yields a Hamilton path in H.   □

4.2 Special Cases and an Upper Bound

Lemma 1. *The minimum size of a guarding set in a complete graph on n vertices is $n - 1$.*

Proof. First note that there always exists a guarding set of size $n-1$ that consists of all edges incident to one vertex.

Now we need to show that any other edge subset of smaller size is not a guarding set. This amounts to stating that every graph with at least $\binom{n}{2} - (n-2)$ edges has a Hamilton path. To see this, proceed by induction on n. Suppose this holds for graphs with less than n vertices. Consider a set S of at most $n - 2$ edges of the complete graph. Let u, v be two vertices with $uv \in S$. One of the two vertices, say v, is incident to at most $\lfloor (n - 2)/2 \rfloor$ edges of S. Consider a Hamilton path on the $n - 1$ vertices $\neq v$, which exists by induction hypothesis. Then vertex v must have one or two incident edges that do not belong to S and connect v to the first, or the last, or two consecutive vertices of this path. Hence we can integrate v in the Hamilton path.   □

Since acyclic orientations of the complete graph K_n correspond to the permutations of S_n, this is in fact the solution to the puzzle "hitting a consecutive pair" in the introduction. The dual graph of the arrangements of hyperplanes corresponding to the complete graph (graphic hyperplane arrangement of K_n) is the skeleton graph of the permutohedron. Hence the above result can also be stated in the following form.

Corollary 1. *The minimum size of a set of zones covering the vertices of the skeleton graph of the n-dimensional permutohedron is $n - 1$.*

We now give a simple, polynomial-time computable upper bound on the size of a guarding set. A set $C \subseteq E$ is an *edge cut* whenever the graph $(V, E \setminus C)$ is not connected.

Theorem 4. *Every edge cut of G is a guarding set of G.*

Proof. Consider an edge cut $C \subseteq E$ and an acyclic orientation A_G of G. This acyclic orientation can be used to define a partial order on V. Let us consider a total ordering of V that extends this partial order, and pick an edge $e = uv \in C$ that minimizes the rank difference between u and v. We claim that e is flippable. Suppose for the sake of contradiction that e is not flippable. Then e must be transitive and there exists a directed path P in A_G between u and v that does not use e. Since C is a cut, u and v belong to distinct connected components of

$(V, E \setminus C)$, and P must use another edge $e' \in C$. By definition, the endpoints of e' have a rank difference that is smaller than that of u and v, contradicting the choice of e.   □

An even shorter proof of the above can be obtained by reusing the following observation from Cordovil and Forge [6]: for every acyclic orientation of G, the set of flippable edges is a spanning set of edges. Therefore, every such set must intersect every edge cut.

While guarding sets are hard to recognize, even for complete graphs, we show that a minimum-size guarding set can be found in polynomial time whenever the input graph is chordal, i.e., if every cycle of length at least 4 has a chord. In that case, the upper bound given by the minimum edge cut is tight, and the result generalizes Lemma 1.

Theorem 5. *The minimum size of a guarding set in a chordal graph is the size of a minimum edge cut.*

Proof. We need to show that whenever a set S of edges of a chordal graph G has size strictly less than the edge connectivity of G, there exists an acyclic orientation of G in which all edges of S are transitive. Let us denote by k the edge connectivity of G, and n its number of vertices.

We proceed by induction on n. For the base case we consider complete graphs. From Lemma 1, the minimum size of a guarding set in K_{k+1} is k. Now suppose the statement holds for every chordal graph with $n - 1$ vertices, and that there exists a k edge connected chordal graph G on n vertices with a guarding set S of size $k - 1$. Since G is chordal, it has at least two nonadjacent simplicial vertices u and v, i.e., vertices whose neighborhood induces a clique. The degree of both u and v is at least k. Hence one of them, say v, is incident to at most $\lfloor (k-1)/2 \rfloor \leq \lfloor (d(v) - 1)/2 \rfloor$ edges of S. Now remove v and consider, using the induction hypothesis, a suitable acyclic orientation of the remaining graph. This orientation induces a total order, i.e., a path p with $d(v)$ vertices, on the neighbors $N(v)$ of v. Then vertex v must have one or two incident edges that do not belong to S and connect v to the first, or the last, or two consecutive vertices of path p. Hence we can integrate v in the path such that all the edges of S that are incident to v are transitive. This yields a suitable acyclic orientation for G and completes the induction step.   □

When the graph is not chordal, it may happen that the minimum size of a guarding set is arbitrarily small compared to the edge connectivity. We can in fact construct a large family of such examples.

Theorem 6. *For every natural number $t \geq 2$ and odd natural number g such that $3 \leq g \leq t$, there exists a graph G with edge connectivity t and a guarding set of size g.*

Proof. The graph G is constructed by considering a wheel with $g + 1$ vertices and center c, and replacing every edge incident to the center c by a copy of the complete graph K_{t-1} such that each vertex of the complete graph is connected to both endpoints of the original wheel edge. Figure 5 shows an example. Let us call C_g the cycle induced by the non-center vertices. Let us first look at the edge connectivity of G. Vertices belonging to one of the K_{t-1} have degree t. On the other hand we easily construct t edge-disjoint paths between every pair of vertices. Hence, the edge connectivity of G equals t.

We now show that the set of edges of C_g is a guarding set.

Suppose the opposite: there exists an acyclic orientation for which no edge of the cycle is flippable, hence all of them are transitive. Since C_g is odd, there must exist two consecutive edges of C_g with the same orientation, say uv and vw, with uv oriented from u to v, and vw from v to w.

The only way to make vw transitive is to construct an oriented path from v to w going through the center c. Hence there must exist an oriented

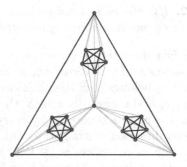

Fig. 5. A graph G with edge connectivity 6 and a guarding set of size 3.

path of the form vPc, where P is a path in the complete graph K_{t-1} attached to v. To make uv transitive we need a directed path $cP'v$. The oriented cycle $vPcP'v$ is in contradiction to the acyclicity. Therefore, some edge of C_g is always flippable. □

5   Median Graphs

A median graph is an undirected graph in which every three vertices x, y, and z have a unique *median*, i.e., a vertex $\mu(x,y,z)$ that belongs to shortest paths between each pair of x, y, and z. Imrich, Klavžar, and Mulder [15] proposed the following construction of median graphs: Start with any triangle-free graph $G = (V, E)$. First add an apex vertex x adjacent to all vertices of V, then subdivide every edge of E once, and let G' be the resulting graph. Figure 6 illustrates the construction.

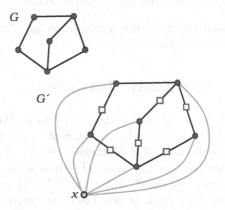

Fig. 6. A graph G and the G' obtained by the construction.

We only need that G' is a partial cube. This can be shown with the explicit construction of an isometric embedding into Q_n. Let $V = \{v_1, \ldots, v_n\}$ and map these vertices bijectively to the standard basis, i.e., $v_i \to e_i$. Apex x is mapped to 0 and the subdivision vertex w_{ij} of an edge $v_i v_j$ is mapped to $e_i + e_j$. The embedding shows that the zones are in one-to-one correspondence with the vertices of G, where the zone of v_i consists of the edge xv_i together with all edges $w_{ij}v_j$. Hence, a zone cover of G' corresponds to a subset of $V(G)$. Let us call w_e the vertex of G' subdividing the edge $e \in E(G)$. The construction of the embedding into Q_n shows that the transformation $G \to G'$ yields a partial cube even if G contains triangles, only membership in the smaller class of median graphs is lost in this case.

Lemma 2. *If G has no isolated vertices, then the minimum size of a zone cover of G' equals the minimum size of a vertex cover of G.*

Proof. We first show that a vertex cover S of G corresponds to a zone cover of G' of the same size. For a vertex w_{ij} at least one of v_i and v_j is in S, hence w_{ij} is covered. If v_i belongs to S then it is covered by its own zone. Otherwise there is an edge v_iv_j and necessarily $v_j \in S$, therefore in this case v_i is covered by the zone of v_j. The apex x is covered by every zone.

The other direction is straightforward. A zone cover of G' must in particular cover all subdivision vertices $w_e \in V(G')$. Hence the zone corresponding to at least one of the endpoint of e must be selected, yielding a vertex cover in G. □

Given a connected triangle-free graph G, one can construct the median graph G' in polynomial time. Since deciding whether G has a vertex cover of size at most k is NP-complete, even on triangle-free graphs, we obtain:

Corollary 2. *Given a median graph G' and an integer k, deciding whether there exists a zone cover of size at most k of G' is NP-complete.*

The minimum vertex cover problem is hard to approximate, even on triangle-free graphs. This directly yields the following corollary.

Corollary 3. *Given a median graph G', finding a minimum size zone cover of G' is APX-hard.*

6   Distributive Lattices

In this special case of the partial cube covering problem, we are given a poset P, and we wish to find a subset S of its elements such that the following holds: for every downset D of P, there exists $x \in S$ such that either $D \cup \{x\}$ or $D \setminus \{x\}$ is a downset, distinct from D. Given a poset P, we refer to a suitable set S as a *guarding set* for P, and let $g(P)$ be the size of a smallest guarding set. Figure 4 gives an example of poset and the corresponding partial cube.

6.1   Relation to Poset Fibres

We first establish a connection between guarding sets and fibres. A *fibre* of a poset is a subset of its elements that meets every nontrivial maximal antichain. Let $f(P)$ be the size of a smallest fibre of P.

Lemma 3. *Every fibre is a guarding set. In particular, $g(P) \leq f(P)$.*

Proof. Consider a poset $P = (V, \leq)$ and one of its downset $D \subseteq V$. Let $F := V \setminus D$, $A := \max(D)$, and $B := \min(F)$, i.e., A is the antichain of maximal elements in the order induced by D and B is the antichain of minimal elements in the order induced by F. By definition, a guarding set is a hitting set for the collection of subsets $A \cup B$ constructed in this way.

Let us now consider the subset $A \cup B' \subseteq A \cup B$, where $B' := \{b \in B : b$ is incomparable to $a, \forall a \in A\}$. This set is easily shown to be a maximal antichain. Hence, it must be hit by any fibre. □

Duffus, Kierstead, and Trotter [8] have shown that every poset on n elements has a fibre of size at most $2n/3$. This directly yields the following.

Corollary 4. *For every n-element poset P, $g(P) \leq 2n/3$.*

We now consider a special case for which the notions of guarding set and fibre coincide. A poset is bipartite if $P = \min(P) \cup \max(P)$, i.e., if the height of P is at most 2.

Lemma 4. *For a bipartite poset P, a set S is a guarding set for P if and only if it is a fibre of P.*

Proof. We know from Lemma 3 that every fibre is a guarding set. The other direction is as follows.

Consider a guarding set S and let A be any maximal antichain of P. Let $T := A \cap \max(P)$, and let D be the downset generated by T. As a downset D is guarded, hence either an element of T is hit by S, or an element of $\min(P \setminus D)$ is hit, but since A is maximal $A = T \cup \min(P \setminus D)$, hence A is hit. Therefore, S is a fibre. □

6.2 Complexity

Lemma 4 yields two interesting corollaries on the complexity of recognizing and finding guarding sets.

Corollary 5. *Given a poset P and a subset S of its elements, the problem of deciding whether S is a guarding set is coNP-complete. This holds even if P is bipartite.*

Proof. Recognition of fibres in bipartite posets has been proved coNP-complete by Duffus et al. [8]. From Lemma 4, this is the same problem as recognizing guarding sets. □

Corollary 6. *Given a poset P and an integer k, the problem of deciding whether there exists a guarding set of size at most k is Σ_2^P-complete. This holds even if P is bipartite.*

Proof. Again, this is a consequence of Lemma 4 and a recent result of Cardinal and Joret [4] showing that the corresponding problem for fibres in bipartite posets is Σ_2^P-complete.   □

Duffus et al. [9] mention that it is possible to construct posets on $15n+2$ elements, every fibre of which must contain at least $8n + 1$ elements, which gives a lower bound with a factor $8/15$. The guarding sets for these examples do not seem to require as many elements.

Open Problems

We left a number of problems open. For instance, we do not know the complexity status of the problem of deciding whether there exists a guarding set of edges of size at most k in a given graph. A natural candidate class would be Σ_2^P. For the same problem, we do not have any nontrivial lower bound on the minimum size of a guarding set. It would also be interesting to give tighter lower and upper bounds on the minimum size of a guarding set in a poset and in a line arrangement. Finally, questions involving partial cubes derived from antimatroids, such as elimination orderings in chordal graphs, are currently under investigation.

Acknowledgments. This work was initiated at the 2nd ComPoSe Workshop held at the TU Graz, Austria, in April 2012. We thank the organizers and the participants for the great working atmosphere. We also acknowledge insightful discussions on related problems with several colleagues, in particular Michael Hoffmann (ETH Zürich) and Ferran Hurtado (UPC Barcelona).

References

1. Ackerman, E., Pach, J., Pinchasi, R., Radoičić, R., Tóth, G.: A note on coloring line arrangements (submitted)
2. Björner, A., Las Vergnas, M., White, N., Sturmfels, B., Ziegler, G.M.: Oriented Matroids. Cambridge University Press, Cambridge (1993)
3. Bose, P., Cardinal, J., Collette, S., Hurtado, F., Korman, M., Langerman, S., Taslakian, P.: Coloring and guarding arrangements. Discrete Math. Theor. Comput. Sci. **15**, 139–154 (2013)
4. Cardinal, J., Joret, G.: Hitting all maximal independent sets of a bipartite graph. Algorithmica (to appear)
5. Cheng, C.T.: A poset-based approach to embedding median graphs in hypercubes and lattices. Order **29**, 147–163 (2012)
6. Cordovil, R., Forge, D.: Flipping in acyclic and strongly connected graphs (submitted)
7. Doignon, J.-P., Falmagne, J.-C.: Learning Spaces - Interdisciplinary Applied Mathematics. Springer, Heidelberg (2011)
8. Duffus, D., Kierstead, H.A., Trotter, W.T.: Fibres and ordered set coloring. J. Comb. Theory Ser. A **58**, 158–164 (1991)
9. Duffus, D., Sands, B., Sauer, N., Woodrow, R.E.: Two-colouring all two-element maximal antichains. J. Comb. Theory Ser. A **57**, 109–116 (1991)

10. Eppstein, D., Falmagne, J.-C., Ovchinnikov, S.: Media Theory - Interdisciplinary Applied Mathematics. Springer, Heidelberg (2008)
11. Fukuda, K., Prodon, A., Sakuma, T.: Notes on acyclic orientations and the shelling lemma. Theor. Comput. Sci. **263**, 9–16 (2001)
12. Füredi, Z.: Maximal independent subsets in Steiner systems and in planar sets. SIAM J. Discrete Math. **4**, 196–199 (1991)
13. Furstenberg, H., Katznelson, Y.: A density version of the Hales-Jewett theorem for $k = 3$. Discrete Math. **75**, 227–241 (1989)
14. Hammack, R., Imrich, W., Klavžar, S.: Handbook of Product Graphs, 2nd edn. CRC Press, Boca Raton (2011)
15. Imrich, W., Klavžar, S., Mulder, H.M.: Median graphs and triangle-free graphs. SIAM J. Discret. Math. **12**, 111–118 (1999)
16. Ovchinnikov, S.: Graphs and Cubes. Springer, New York (2011)
17. Stanley, R.P.: Enumerative Combinatorics, vol. 1. Cambridge University Press, Cambridge (1997)

The Non-confusing Travel Groupoids
on a Finite Connected Graph

Jung Rae Cho[1], Jeongmi Park[1(✉)], and Yoshio Sano[2]

[1] Department of Mathematics, Pusan National University, Busan 609-735, Korea
{jungcho,jm1015}@pusan.ac.kr
[2] Division of Information Engineering, Faculty of Engineering,
Information and Systems, University of Tsukuba, Ibaraki 305-8573, Japan
sano@cs.tsukuba.ac.jp

Abstract. The notion of travel groupoids was introduced by L. Nebeský in 2006 in connection with a study on geodetic graphs. A travel groupoid is a pair of a set V and a binary operation $*$ on V satisfying two axioms. For a travel groupoid, we can associate a graph. We say that a graph G has a travel groupoid if the graph associated with the travel groupoid is equal to G. A travel groupoid is said to be non-confusing if it has no confusing pairs. Nebeský showed that every finite connected graph has at least one non-confusing travel groupoid.

In this note, we study non-confusing travel groupoids on a given finite connected graph and we give a one-to-one correspondence between the set of all non-confusing travel groupoids on a finite connected graph and a combinatorial structure in terms of the given graph.

Keywords: Travel groupoid · Confusing pair · Non-confusing travel groupoid · Geodetic graph · Spanning tree

2010 Mathematics Subject Classification: 20N02, 05C12, 05C05.

1 Introduction

A *groupoid* is the pair $(V, *)$ of a nonempty set V and a binary operation $*$ on V. The notion of travel groupoids was introduced by L. Nebeský [5] in 2006 in connection with his study on geodetic graphs [1–3] and signpost systems [4]. First, let us recall the definition of travel groupoids.

A *travel groupoid* is a groupoid $(V, *)$ satisfying the following axioms (t1) and (t2):

(t1) $(u * v) * u = u$ (for all $u, v \in V$),
(t2) if $(u * v) * v = u$, then $u = v$ (for all $u, v \in V$).

Jung Rae Cho—This research was supported for two years by Pusan National University Research Grant.
Yoshio Sano—This work was supported by JSPS KAKENHI grant number 25887007.

© Springer International Publishing Switzerland 2014
J. Akiyama et al. (Eds.): JCDCGG 2013, LNCS 8845, pp. 14–17, 2014.
DOI: 10.1007/978-3-319-13287-7_2

A *geodetic graph* is a connected graph in which there exists a unique shortest path between any two vertices. Let G be a geodetic graph, and let $V := V(G)$. For two vertices u and v of G, let $A_G(u, v)$ denote the vertex adjacent to u which is on the unique shortest path from u to v in G. Define a binary operation $*$ on V as follows: For all $u, v \in V$, let $u * v := A_G(u, v)$ if $u \neq v$ and $u * v := u$ if $u = v$. This groupoid $(V, *)$ is called the *proper groupoid* of the geodetic graph G. Remark that the proper groupoid of any geodetic graph is a travel groupoid.

Let $(V, *)$ be a travel groupoid, and let G be a graph. We say that $(V, *)$ *is on G* or that *G has $(V, *)$* if $V(G) = V$ and $E(G) = \{\{u, v\} \mid u, v \in V, u \neq v,$ and $u * v = v\}$. Note that every travel groupoid is on exactly one graph.

Let $(V, *)$ be a travel groupoid. For $u, v \in V$, we define $u *^0 v := u$ and $u *^{i+1} v := (u *^i v) * v$ for every nonnegative integer i. It is clear that $(u *^j v) *^k v = u *^{j+k} v$ holds for any nonnegative integers j and k. Note that, for any two distinct elements u and v in V, it holds that $u * v \neq u$ (see [5, Proposition 2 (2)]) and that $u *^2 v \neq u$ by (t2). An ordered pair (u, v) of two distinct elements of V is called a *confusing pair* in $(V, *)$ if there exists an integer $i \geq 3$ such that $u *^i v = u$. A travel groupoid $(V, *)$ is said to be *non-confusing* if there is no confusing pair in $(V, *)$.

Nebeský gave a characterization of non-confusing travel groupoids on a finite graph.

Theorem 1 ([5, Theorem 3]). *Let $(V, *)$ be a travel groupoid on a finite graph G. Then, $(V, *)$ is non-confusing if and only if, for all distinct elements u and v in V, there exists a positive integer k such that the sequence $(u *^0 v, \ldots, u *^{k-1} v, u *^k v)$ is a path from u to v in G.* □

Nebeský also showed a result on the existence of non-confusing travel groupoids. Note that a travel groupoid $(V, *)$ is said to be *simple* if it satisfies the following axiom (t3) if $v * u \neq u$, then $u * (v * u) = u * v$ (for all $u, v \in V$).

Theorem 2 ([5, Theorem 4]). *For every finite connected graph G, there exists a simple non-confusing travel groupoid on G.* □

By Theorem 2, we know that every finite connected graph has always at least one non-confusing travel groupoid.

In this note, we study non-confusing travel groupoids on a given finite connected graph and we give a one-to-one correspondence between the set of all non-confusing travel groupoids on a finite connected graph and a combinatorial structure in terms of the given graph.

2   Main Result

To describe the structure of non-confusing travel groupoids on a graph, we define the following.

Definition. For a vertex v of a graph G, a *v-tree* is a spanning tree of G which contains all the edges incident to v. We denote by $\mathcal{S}_G(v)$ the set of all v-trees in G. □

Lemma 3. *Let* $(V, *)$ *be a non-confusing travel groupoid on a finite connected graph* G. *For an element* v *of* V, *let* T_v *be the graph defined by*

$$V(T_v) := V \quad and \quad E(T_v) := \{\{u, u * v\} \mid u \in V \setminus \{v\}\}.$$

Then, T_v *is a* v-*tree of* G.

Proof. Since $(V, *)$ is on G, for any two distinct elements u and v in V, $\{u, u*v\}$ is an edge of G. Therefore, we have $E(T_v) \subseteq E(G)$. Moreover, $V(T_v) = V = V(G)$. So T_v is a spanning subgraph of G. By the definition of the graph T_v, we have $|E(T_v)| \leq |V| - 1$. Since $(V, *)$ is non-confusing, it follows from Theorem 1 that any vertex u of T_v distinct from v is connected to the vertex v by a path P_{uv} in G, where P_{uv} is of the form $(u *^0 v, \ldots, u *^k v)$ for some positive integer k. Since $u *^{i+1} v = (u *^i v) * v$, the edge $\{u *^i v, u *^{i+1} v\}$ of the path P_{uv} is also an edge of T_v for any $i \in \{0, 1, \ldots, k-1\}$. Therefore, the path P_{uv} from u to v in G is also a path in T_v. So we can conclude that T_v is connected. Since T_v is a connected graph with $|E(T_v)| \leq |V(T_v)| - 1$, T_v is a tree. Thus, T_v is a spanning tree of G. For each vertex x which is adjacent to the vertex v in G, we have $x * v = v$ and therefore $\{x, v\} \in E(T_v)$. So, all the edges incident to the vertex v in G are contained in T_v. Hence, the graph T_v is a v-tree of G. □

The following theorem is our main result.

Theorem 4. *Let* G *be a finite connected graph. Then, there exists a one-to-one correspondence between the set* $\Pi_{v \in V(G)} \mathcal{S}_G(v)$ *and the set of all non-confusing travel groupoids on* G.

Proof. Let $V := V(G)$, and let $\mathcal{N}_G$ be the set of all non-confusing travel groupoids on G. Note that, since G is finite, both the sets $\Pi_{v \in V} \mathcal{S}_G(v)$ and $\mathcal{N}_G$ are finite.

For any $(T_v)_{v \in V} \in \Pi_{v \in V(G)} \mathcal{S}_G(v)$, we define a groupoid on V as follows: For two distinct elements u and v in V, $u * v$ is defined to be the vertex adjacent to u which is on the unique path from u to v in the tree T_v. If $u = v$, then let $u * v = u$. We show that this groupoid $(V, *)$ is a non-confusing travel groupoid. First, we check that $(V, *)$ satisfies the axioms (t1) and (t2). Take any two distinct elements u and v in V. Let $w := u * v$. Then $\{u, w\} \in E(G)$. Therefore $\{w, u\} \in E(T_u)$ and we have $(u * v) * u = w * u = u$. Moreover, if $u = v$, then $(u * v) * u = (u * u) * u = u * u = u$. Thus the axiom (t1) holds. Again, take any two distinct elements u and v in V. If u and v are adjacent (i.e. $\{u, v\} \in E(G)$), then $\{u, v\} \in E(T_v)$ and therefore $(u * v) * v = v * v = v \neq u$. If u and v are not adjacent, then $u * v \neq v$ and the element $(u * v) * v$ is the third vertex of the path from u to v in the tree T_v and therefore $(u * v) * v$ is not the element u. Thus the axiom (t2) holds. So, $(V, *)$ is a travel groupoid. Second, we check that $(V, *)$ is non-confusing. Take any two distinct elements u and v in V. Let k be the length of the path from u to v in T_v. Then, it follows from the definition of

$(V, *)$ that $u *^k v = v$ and that $(u *^0 v, \ldots, u *^k v)$ is the path from u to v in T_v. Therefore, $(u*^0 v, \ldots, u*^k v)$ is a path from u to v in G. By Theorem 1, the travel groupoid $(V, *)$ is non-confusing. Now, we define a map $\Phi : \Pi_{v \in V} \mathcal{S}_G(v) \to \mathcal{N}_G$ by $\Phi((T_v)_{v \in V}) = (V, *)$, where $(V, *)$ is the non-confusing travel groupoid defined as above for $(T_v)_{v \in V}$.

Next, let $(V, *)$ be a non-confusing travel groupoid on a finite connected graph G. For each $v \in V(G)$, we define a graph T_v by $V(T_v) := V$ and $E(T_v) := \{\{u, u * v\} \mid u \in V \setminus \{v\}\}$. By Lemma 3, T_v is a v-tree of G, i.e., $T_v \in \mathcal{S}_G(v)$. Therefore, $(T_v)_{v \in V} \in \Pi_{v \in V} \mathcal{S}_G(v)$. Now, we define a map $\Psi : \mathcal{N}_G \to \Pi_{v \in V} \mathcal{S}_G(v)$ by $\Psi((V, *)) = (T_v)_{v \in V}$, where $(T_v)_{v \in V}$ are v-trees defined as above for $(V, *)$.

Then, we can check that $\Psi \circ \Phi((T_v)_{v \in V}) = (T_v)_{v \in V}$ holds for any $(T_v)_{v \in V} \in \Pi_{v \in V} \mathcal{S}_G(v)$ and that $\Phi \circ \Psi((V, *)) = (V, *)$ holds for any $(V, *) \in \mathcal{N}_G$. Hence the map Φ is a one-to-one correspondence between the sets $\Pi_{v \in V} \mathcal{S}_G(v)$ and $\mathcal{N}_G$. □

Corollary 5. *Let G be a finite connected graph. Then, the number of non-confusing travel groupoids on G is equal to $\Pi_{v \in V(G)} |\mathcal{S}_G(v)|$.*

Proof. It follows from Theorem 4. □

References

1. Nebeský, L.: An algebraic characterization of geodetic graphs. Czech. Math. J. **48**(123), 701–710 (1998)
2. Nebeský, L.: A tree as a finite nonempty set with a binary operation. Mathematica Bohemica **125**, 455–458 (2000)
3. Nebeský, L.: New proof of a characterization of geodetic graphs. Czech. Math. J. **52**(127), 33–39 (2002)
4. Nebeský, L.: On signpost systems and connected graphs. Czech. Math. J. **55**(130), 283–293 (2005)
5. Nebeský, L.: Travel groupoids. Czech. Math. J. **56**(131), 659–675 (2006)

Decomposing Octilinear Polygons
into Triangles and Rectangles

Serafino Cicerone[⊠] and Gabriele Di Stefano

Dipartimento di Ingegneria e Scienze dell'Informazione e Matematica,
Università degli Studi dell'Aquila, L'Aquila, Italy
{serafino.cicerone,gabriele.distefano}@univaq.it

Abstract. In this paper we study the minimal decomposition of octilinear polygons with holes into octilinear triangles and rectangles. This new problem is relevant in the context of modern electronic CAD systems, where it arises when the generation and propagation of electromagnetic noise into multi-layer PCBs has to be detected. It can be seen as a generalization of a problem deeply investigated in the last decades: the minimal decomposition of rectilinear polygons into rectangles, which is solvable in polynomial time. We show that the new problem is NP-hard. We also show the NP-hardness of a related problem, that is the decomposition of an octilinear polygon with holes into octilinear convex polygons. For both problems, we propose efficient approximation algorithms.

Keywords: Polygon decomposition · Octilinear polygons · Approximation algorithms · CAD applications

1 Introduction

The problems of partitioning a planar polygon into simpler components are well studied in computational geometry, as they have various applications in computer graphics, mesh generation, pattern recognition, VLSI, and more. Particular examples of these problems are the decomposition of polygons into convex components, triangles, or other exotic shapes like spiral, star-shaped, and monotone components. (e.g., see [10] and references therein).

A specific problem concerns the partition into a minimum number of rectangles of a rectilinear polygon, that is a polygon having the edges parallel to two orthogonal directions [16]. Despite its specificity, rectangular partition have many applications in VLSI, DNA micro-array design, processing and compression of bitmap images [2,7]. The rectilinear polygon decomposition problem has been solved in polynomial-time at earlier stage; for instance, we can find a

Work supported by the Research Grant 2012C4E3KT "PRIN 2012" Amanda (Algorithmics for MAssive and Networked DAta) from the Italian Ministry of University and Research.

© Springer International Publishing Switzerland 2014
J. Akiyama et al. (Eds.): JCDCGG 2013, LNCS 8845, pp. 18–30, 2014.
DOI: 10.1007/978-3-319-13287-7_3

$O(n^{2.5})$ time algorithm in [8, 14] and a $O(n^{1.5} \log n)$ time algorithms in [9, 13]. In the last decades many works have been devoted to specific instances of this problem (e.g., see [5, 6]).

In this paper, motivated by a renewed interest in the VLSI field, we present a generalization of the rectilinear polygon decomposition problem. This new problem is motivated as follows.

Motivation. Currently, microprocessors and application-specific integrated circuits have thousands of gates switching simultaneously. The impulsive and repetitive current drawn by these active devices from the power delivery network (PDN) is a challenging issue for a correct and reliable PDN design and a severe source of electromagnetic noise generation [17]. The PDN for modern medium-to-high-speed digital printed circuit boards (PCBs) is usually formed from one or more pair of conducting parallel planes used as "power" and "ground". The PDN for digital circuitry has evolved over time, as signal and clock speeds have increased, from discrete power supply wires, to discrete traces, to area fills and ground islands on single/two-layer slow-speed boards, to the planar power bus structure used extensively in today's multi-layer high-speed PCBs.

Noise generated in the power bus can be easily propagated throughout the board. Propagated noise can affect the operation of other active devices (signal integrity) [15]. Among the possible techniques to study the generation and propagation of noise there is the so-called Cavity Model [11] in which facing portions of power bus are considered electromagnetic resonant cavities. Given a real-world board's layout, one of the primary requirements for the application of this technique is the geometric identification of all the cavities and their connectivity. Then, a suitable processing of the geometrical cavities' boundaries is requested for a correct and not over-detailed electromagnetic modeling. After these actions, the geometry dataset (containing also the electrical parameters) is ready for being input to the *cavity model solver*.

From a geometrical point of view, a power bus corresponds to a simple polygon with holes. Two facing polygons P_1 and P_2 located at parallel layers L_1 and L_2 form a cavity; this cavity is geometrically defined as a polyhedron having the polygon $P = P_1 \cap P_2$ as base area and the distance from L_1 and L_2 as height. The currently available cavity model solvers require that such a polygon P must be either a *rectangle and isosceles triangle* or a *rectangle*. This and other constraints lead to solve the problem of computing cavities [4] according to the following two steps:

1. "simplify" each base polygon P into an octilinear polygon P' (a polygon is octilinear when its angles are all multiple of $45°$). As usual, this simplification must be performed according to some error criterion;
2. decompose the octilinear polygon P' into octilinear triangles and rectangles.

The problem at the former step as been studied in [3]. In this work we address the polygon decomposition problem at the latter step.

Results. The main aim of this paper is to study the decomposition of octilinear polygons with holes into octilinear triangles and rectangles. We call this problem

OPD-TR (*Octilinear Polygon Decomposition into octilinear Triangles and Rectangles*). After providing some preliminary results, we study the computational complexity of the OPD-TR problem and we show that it is NP-hard. Then, we prove that OPD-TR is in APX by providing an $O(n \log n)$-time 16-approximation algorithm. We also provide an $O(n \log n)$-time 3-approximation algorithm for OPD-TR when the input is restricted to a subclass of octilinear polygons. The approximation algorithms have been achieved by exploiting results on a strictly related problem, that is the *Octilinear Polygon Decomposition into octilinear Convex Components* (OPD-C), which also has theoretical interest in itself. We show that this problem is NP-hard and we provide an $O(n \log n)$-time exact algorithm for the decomposition of a subclass of octilinear polygons.

Due to lack of space, many proofs are omitted (they will be given in the full version).

2   Definitions and Problem Statements

In this paper we consider the following types of polygonal objects:

- a *polygon* P is a compact region of the plane, bounded by a simple closed polygonal line. A polygon may also have *holes*, that is internal pairwise-disjoint polygonal lines delimiting regions of the plane not belonging to the polygon itself;
- a *component* is a polygon without holes;
- a *multiple polygon* $\mathcal{P}$ is a set of polygons. The intersection of any two elements of $\mathcal{P}$, if non-empty, consists totally of edges and vertices.

The *boundary* of a polygon P consists of all points of the polygonal lines defining P and its holes, while the *interior* of P consists of all points of P that do not belong to the boundary of P.

A polygonal line can be represented by a finite sequence of vertices $v_1, v_2, \ldots, v_n$ and edges $[v_1, v_2], [v_2, v_3], \ldots, [v_{n-1}, v_n], [v_n, v_1]$ such that two edges share at most one point and this point is shared only if the edges are consecutive.

Let us consider non-oriented directions in the plane, and define them according to the angle they form with respect to the x-axis. Directions forming angles multiples of 90° are called *rectilinear*, while those forming angles multiples of 45° are called *octilinear*. Notice that, there are two rectilinear and four octilinear directions. A polygon edge parallel to an octilinear (rectilinear, resp.) direction is called octilinear edge (rectilinear edge, resp.).

Definition 1. *An octilinear polygon (rectilinear polygon, resp.) is a polygon whose edges are all octilinear (rectilinear, resp.).*

Figure 1 shows internal angles of octilinear and rectilinear polygons. We call the angles of 45°, 90°, 135°, 225°, 270°, 315° being of type a, b, c, d, e and f, respectively (and, we extend the notion of *type* to vertices: the type of v_i is the type of the angle at v_i). Without loss of generality, we do not consider polygons

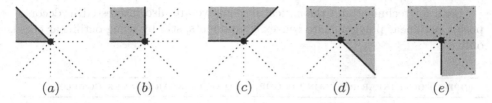

(a)   (b)   (c)   (d)   (e)

Fig. 1. All the possible internal angles at vertices in octilinear polygons (angles d, e and f are concave). Angles b and e are the only possible internal angles of rectilinear polygons.

with internal angles of 180°. Given a multiple polygon $\mathcal{P}$, we denote by $A^{\mathcal{P}}$, (resp., $B^{\mathcal{P}}, C^{\mathcal{P}}, D^{\mathcal{P}}, E^{\mathcal{P}}, F^{\mathcal{P}}$) the number of angles of type a (resp., $b, c, d, e, f,$) in $\mathcal{P}$. We write A, instead of $A^{\mathcal{P}}$, if $\mathcal{P}$ is clear by the context, and we do the same for the other types of angle. Note that angles of type a, b, c (resp., d, e, f) are convex (resp., concave). A component with only convex angles is called *convex component*.

We use *cuts* to divide polygons. A cut for a polygon P can be defined as a segment $c = [p_1, p_2]$ such that both p_1 and p_2 belong to the boundary of P and each internal point of c belongs to the interior of P. A cut parallel to a rectilinear (octilinear, resp.) direction is called *rectilinear cut* (*octilinear cut*, resp.). Figure 2a shows an octilinear polygon P with two rectilinear cuts that divide P into two triangles and one rectangle.

A *decomposition* of a polygon P is defined by a sequence $(c_1, c_2, \ldots, c_m)$ of cuts for P that, when applied in order, produces a multiple polygon $\mathcal{P} = \{P_1, P_2, \ldots, P_{m'}\}$ such that: (1) the union of elements of $\mathcal{P}$ gives P, and (2) the intersection of any two elements of $\mathcal{P}$, if non-empty, consists totally of edges and vertices. In general, we are interested in decomposition producing multiple polygons having only components as elements. Generalizing, decomposing a multiple polygon $\mathcal{P}$ produces a new multiple polygon $\mathcal{P}'$ consisting of the union of the decompositions of all elements of $\mathcal{P}$.

In this work we are interested in decomposing octilinear polygons into octilinear triangles and rectangles. In the remainder, octilinear rectangles and triangles are simply called *basic components*.

(OPD-TR) OCTILINEAR POL. DECOMP. INTO OCTILINEAR TRIANGLES AND RECTANGLES

GIVEN: A multiple octilinear polygon $\mathcal{P}$.
PROBLEM: Decompose $\mathcal{P}$ into the minimum number of basic components.

When the input of the OPD-TR problem is restricted to rectilinear polygons, it can be easily observed that we get the well known problem of "minimum dissection of rectilinear regions" [16], that has been solved in polynomial-time at earlier stage.

Both for technical and theoretical reasons we are also interested in decomposing octilinear polygons into convex components, still by using octilinear cuts only.

(OPD-C) OCTILINEAR POLYGON DECOMP. INTO OCTILINEAR CONVEX COMPONENTS

GIVEN: A multiple octilinear polygon $\mathcal{P}$.
PROBLEM: Decompose $\mathcal{P}$ into the minimum number of octilinear convex components.

---

3   Preliminary Results

As observed in the previous section, the OPD-TR problem is a generalization of the well known problem of "minimum dissection of rectilinear regions" [16], denoted here as OPD-TR$^-$. As already mentioned, this problem is solvable in polynomial time. Why this problem is so easy? On the sight of addressing the OPD-TR problem, it is useful to provide a brief explanation.

Concerning the OPD-TR$^-$ problem, note that a rectilinear polygon has vertices of type b and e only, while rectangles (i.e., the final components of the decomposition) has vertices b only. This implies that, solving the OPD-TR$^-$ problem requires to use cuts that "remove" angles of type e, i.e. all the concave vertices. To remove the concave vertices, two kind of cuts can be used:

- **vertex-cut:** a vertex-cut is a rectilinear cut $[v, v']$, where v is the concave vertex to be removed and v' is any other vertex. At a closer analysis, in the case of rectilinear polygons, v' must be a concave vertex as well.
- **point-cut:** a point-cut is a rectilinear cut $[v, p]$, where v is the concave vertex to be removed and p is a point internal to some edge (an original edge, or some edge generated by cuts).

In [7] it is recalled that the OPD-TR$^-$ problem can be efficiently solved by using a two-phase approach: first find and apply a maximum set of disjoint rectilinear vertex-cuts, then for each concave vertex v (taken in an arbitrary order) apply a rectilinear point-cut $[v, p]$. Now we could ask whether such an approach can be useful to face with the general case, i.e. when the input is an octilinear polygon. Unfortunately, the answer is negative since at least the following observations apply:

- for the rectilinear case, when a point-cut $[v, p]$ that eliminates the concavity of v meets a point p (where p is a point internal to some edge), the cut *does not create* new concave vertices (in fact, it creates two new convex vertices, i.e. both of type b).
  In the octilinear case, a cut $[v, p]$ may create at p two new vertices of type a and c (cf polygons in Fig. 2). Hence, the new concave vertex of type c needs to be further removed by planning a new octilinear cut, and this new cut may create further concave vertices and so on.

– for the rectilinear case, the presence of vertex-cuts represents a noteworthy situation, since they belong to the solution. In the octilinear case this is not true, as shown by Fig. 2b, where the vertex-cut $[v_1, v_2]$ does not belong to any minimal solution.

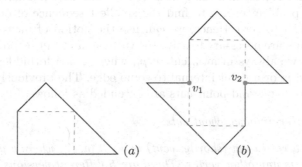

Fig. 2. Examples of decomposition of an octilinear polygon. The vertex-cut $[v_1, v_2]$ does not belong to the solution of the OPD-TR problem.

3.1 Cuts for the OPD-TR Problem

Concerning the OPD-TR problem, note that a polygon is a basic component if and only if its internal angles are of type a and b. This implies that, solving the OPD-TR problem requires to use octilinear cuts that remove vertices of type c, d, e and f. Similarly, for the OPD-C problem, octilinear cuts must remove vertices of type d, e and f. All the vertices to be removed can be considered as "forbidden" in the corresponding problem.

Definition 2. *Vertices (or angles) of type c, d, e and f are forbidden for the OPD-TR problem. Vertices (or angles) of type d, e and f are forbidden for the OPD-C problem.*

We simply use the term *forbidden* when the problem is clear or when we have properties that hold for forbidden vertices/angles independently from the problem at hand.

Definition 3. *Let v a vertex of an octilinear polygon P. The measure of v in P corresponds to the minimum number of octilinear cuts required to divide the internal angle in v into subangles of type non-forbidden or $180°$. The measure of v in P is denoted by $\mu_P(v)$.*

Concerning the OPD-TR problem, by definition, the measure of vertices of type a, b, c, d, e, and f are 0, 0, 1, 1, 1, and 2, respectively. Concerning the OPD-C problem, the measure of vertices of type a, b, c, d, e, and f are 0, 0, 0, 1, 1, and

1, respectively. Note that, in both cases forbidden vertices have measure greater than zero.

The *total measure* of an octilinear polygon P with vertices $v_1, v_2, \ldots, v_n$ is defined as $\mu(P) = \sum_i \mu_P(v_i)$, and the *total measure* of a multiple octilinear polygon $\mathcal{P}$ with polygons $P_1, P_2, \ldots, P_n$ is defined as $\mu(\mathcal{P}) = \sum_i \mu(P_i)$.

According to the notion of total measure we could say that solving the OPD-TR (or OPD-C) problem means to find the smallest sequence of octilinear cuts that reduces $\mu(\mathcal{P})$ to zero. Hence, we can use the notion of measure to define which kind of cuts are necessary to solve the OPD-TR and OPD-C problems. From now on, by *cut* we always mean a cut $[v, p]$, where v is a forbidden vertex and p is either a vertex or a point internal to some edge. The previously introduced concepts of vertex-cuts and point-cuts are extended as follows.

Definition 4. *Terminology about cuts:*

– **vertex-cut:** *a vertex-cut (shortly, vcut) is a cut $[v, v']$, where v is a forbidden vertex and v' is any other vertex. There are 5 different versions of vcuts:*
  – $\langle n \rangle$-*vcut, where $-2 \leq n \leq 2$ represents how much the overall measure of an octilinear polygon decreases by applying the cut.*
  A forbidden-vertex-cut (shortly, fvcut) is a cut $[v, v']$, where both v and v' are forbidden.
– **point-cut:** *a point-cut (shortly, pcut) is a cut $[v, p]$, where v is a forbidden vertex and p is a point internal to some edge. There are 2 different versions of pcuts:*
  – $\langle b, b \rangle$-*pcut, if the cut generates at p two angles of type b;*
  – $\langle a, c \rangle$-*pcut, if the cut generates at p angles of type a and c.*

Lemma 1. [16] *Solutions to the OPD-TR$^-$ problem can be found by using $\langle 2 \rangle$-vcut or $\langle b, b \rangle$-pcut only.*

Lemma 2. *Solutions to the OPD-TR problem may require the following cuts: $\langle n \rangle$-vcut, $-1 \leq n \leq 2$, $\langle b, b \rangle$-pcut or $\langle a, c \rangle$-pcut.*

It remains open to check whether $\langle -2 \rangle$-vcuts are necessary to solve the OPD-TR problem.

Definition 5. *Octilinear (multiple) polygons having no fvcuts are called non-degenerated (multiple) polygons.*

The following lemma provides a useful result which is valid for both the OPD-TR and OPD-C problems.

Lemma 3. *Let $\mathcal{P}$ be a non-degenerated multiple polygon. If there exists a decomposition into basic components (convex components, respectively) that requires $\mu(\mathcal{P})$ cuts, then it is an optimal solution for the OPD-TR problem (for the OPD-C problem, respectively).*

Proof. By definition, $\mu(P) = \sum_i \mu_P(v_i)$, where $\mu_P(v_i)$ is the minimum number of octilinear cuts required to divide the internal angle in v_i into non-forbidden subangles. Hence, each forbidden vertex v_i contribute to $\mu(P)$ by adding 1 or 2. This implies that decomposing $\mathcal{P}$ by using less than $\mu(\mathcal{P})$ cuts requires the existence of $\langle 2 \rangle$-vcuts, i.e. cuts connecting pairs of forbidden vertices. This is not the case, since the input multiple polygon $\mathcal{P}$ is non-degenerated. $\square$

Notice that the polygon shown in Fig. 2a needs more than $\mu(P) = 1$ cuts. We close the section by observing that all the four octilinear directions may be necessary to get the optimal solution to the OPD-TR problem.

Lemma 4. *Concerning the* OPD-TR *problem, the following properties hold:*

1. *an optimal solution may require cuts parallel to all the four octilinear directions;*
2. *without using all the four octilinear directions may lead to solutions that are* k *times the optimal one, with* k *arbitrarily large.*

4   Complexity of the Main Problem

The computational complexity of the OPD-TR problem is stated in Theorem 1. For the sake of space, we omit the long proof of this theorem which is borrowed from [12], where the problem of decomposing a polygon into a minimum number of convex components by cuts in the directions of $\mathcal{F}$ ($\mathcal{F}$ is a family of non-oriented directions in the plane) is studied. There, authors have shown that the problem is NP-hard if $|\mathcal{F}| \geq 3$ and is solvable in polynomial time if $|\mathcal{F}| \leq 2$.

Theorem 1. *The* OPD-TR *problem is NP-hard.*

5   Decomposition into Convex Components

In this section we study the OPD-C problem: we start by stating the computational complexity, and then we provide an optimal algorithm when the input is restricted to non-degenerated octilinear polygons. This algorithm will be used to achieve approximation results shown in the next section. We conclude this section by providing a lemma for multiple octilinear polygons that extends a property found for rectilinear polygons.

Theorem 2. *The* OPD-C *problem is NP-hard.*

Theorem 3. *There exists an* $O(n \log n)$-*time algorithm for solving the* OPD-C *problem restricted to non-degenerated octilinear polygons.*

Proof. Given a non-degenerated octilinear polygon P, according to Lemma 3, we prove the theorem by showing that there exists a solution for the OPD-C problem that requires $\mu(P)$ octilinear cuts. We decompose P according to the following approach: *for each concave vertex* v, *perform a cut* $[v, x]$ *that prolongs an edge incident to* v. To show that such an approach produces a solution for the OPD-C problem, consider two different cases, according to x:

1. x is a point p interior to some edge e of P. Hence the cut $[v, x]$ is either a $\langle b, b \rangle$-pcut or a $\langle a, c \rangle$-pcut. In both cases, the new angles at p are convex and then they do not need further cuts. So, one cut is sufficient to "remove" the concave vertex v from P.
2. x coincides with a vertex v' of P, but v' is not forbidden. Also in this case no further cuts are needed to "remove" the concave vertex v from P.

This proves that the above approach uses $\mu(P)$ octilinear cuts exactly. Concerning the execution time, the algorithm performs $O(n)$ cuts and each cut can be performed in $O(\log n)$-time by using the ray-shooting algorithm given in [1]. It is worth to note that, if we need to test whether the input polygon is indeed a non-degenerated component, then this task can be easily accomplished in $O(n \log n)$-time by using a brute-force approach: test whether for each forbidden vertex v and for each ray r (shoot from v along an octilinear direction d), the first point reached by r is a forbidden vertex or not.   □

In [16] it has been shown that given a multiple rectilinear polygon $\mathcal{P}$ consisting of k polygons and w holes, then $B^{\mathcal{P}} - E^{\mathcal{P}} = 4k - 4w$. A generalization of this result is given by the following lemma.

Lemma 5. *Given a multiple octilinear polygon $\mathcal{P}$ consisting of k polygons and w holes, then $3A^{\mathcal{P}} + 2B^{\mathcal{P}} + C^{\mathcal{P}} - D^{\mathcal{P}} - 2E^{\mathcal{P}} - 3F^{\mathcal{P}} = 8k - 8w$.*

Since a convex component P has internal angles of type a, b, c only, then, by Lemma 5, we have $3A^P + 2B^P + C^P = 8$. As consequence, the number of convex components which differ by the (circular) sequence of convex angles is finite.

6   Approximation Results

We start by providing an approximation result for the OPD-TR problem restricted to non-degenerated components. To this end we need the notion of *nice-cut*:

Definition 6. *A cut in an octilinear polygon is called* nice-cut *if either it is a $\langle b, b \rangle$-pcut, or it is a $\langle +2 \rangle$-vcut.*

Lemma 6. *Let P be a convex component with at least 5 vertices. Then, there exists a nice-cut in P.*

Algorithm 1 provides an approximation for the OPD-TR problem restricted to non-degenerated components. This algorithm takes as input a non-degenerated component P (i.e., an octilinear polygon without holes and without forbidden-vertex-cuts) and performs, as a preliminary phase, the decomposition of P into octilinear convex components. For this preliminary phase, the optimal decomposition algorithm presented in the proof of Theorem 3 is used. The following lemmas and remarks are useful to state the approximation bound of Algorithm 1.

Lemma 7. *Let $P = \{v_1, v_2, \ldots, v_n\}$ be a non-degenerated component. Then, each solution for P requires at least $\mu(P) = C^P + D^P + E^P + 2F^P$ cuts.*

Algorithm 1. Decomposing a non-degenerated component

Data: a non-degenerated component $P = \{v_1, v_2, \ldots, v_n\}$
Result: a decomposition of P into basic components

1 **foreach** *vertex v_i of type d, e and f* **do**
2   divide the angle at v_i by a cut that prolongs an edge incident to v_i ;
    /* according to the proof of Theorem 3, at the end of this step,
        P is optimally decomposed into convex components   */
3 **end**
4 **while** *there are convex components with at least 5 vertices* **do**
5   take a convex component C with at least 5 vertices;
6   **foreach** *vertex v_i of type c* **do**
7   decompose C by using a nice-cut $[v_i, p]$;
       /* according to Lemma 6, there exists such a cut in C */
8   **end**
9 **end**
10 **foreach** *convex component C with 4 vertices and different from a rectangle* **do**
11   decompose C ;
12 **end**
13 **return** *the set of all the obtained basic components*

Remark 1. Lemma 7 can be rephrased by stating that each solution for a non-degenerated component P requires at least $C^P + D^P + E^P + 2F^P + 1$ basic components. It is possible to extend easily this result to any multiple octilinear polygon $\mathcal{P}$ with k polygons and w holes: in such a case, at least $(C^{\mathcal{P}} + D^{\mathcal{P}} + E^{\mathcal{P}} + 2F^{\mathcal{P}})/2 + k - w$ basic components are required. The ratio $(C^{\mathcal{P}} + D^{\mathcal{P}} + E^{\mathcal{P}} + 2F^{\mathcal{P}})/2$ is due to possible forbidden-vertex-cuts (where a cut may connect two forbidden vertices).

Lemma 8. *Let P be a non-degenerated component. Algorithm 1 decomposes P by using at most $2C^P + 3D^P + 3E^P + 5F^P$ cuts.*

Remark 2. Lemma 8 can be rephrased by stating that Algorithm 1 decomposes P by producing at most $2C^P + 3D^P + 3E^P + 5F^P + 1$ basic components. This result can be extended to any multiple octilinear polygon $\mathcal{P}$ with k polygons and w holes: in such a case Algorithm 1 produces at most $2C^{\mathcal{P}} + 3D^{\mathcal{P}} + 3E^{\mathcal{P}} + 5F^{\mathcal{P}} + k - w$ basic components.

Theorem 4. *There exists an $O(n \log n)$-time 3-approximation algorithm for the* OPD-TR *problem when the input is restricted to non-degenerated components.*

According to Remarks 1 and 2, we know that Algorithm 1 is able to process not only non-degenerated components but also generic multiple octilinear polygons. This implies that we can provide an approximation result for the OPD-TR problem, as follows.

Theorem 5. *There exists an $O(n \log n)$-time 16-approximation algorithm for the* OPD-TR *problem.*

Proof. Let us assume that a multiple octilinear polygon $\mathcal{P}$ consisting of k polygons and w holes is given as input to Algorithm 1. According to Remarks 1 and 2 we have to show that

$$\frac{5F^{\mathcal{P}} + 3E^{\mathcal{P}} + 3D^{\mathcal{P}} + 2C^{\mathcal{P}} + k - w}{(C^{\mathcal{P}} + D^{\mathcal{P}} + E^{\mathcal{P}} + 2F^{\mathcal{P}})/2 + k - w} \leq 16.$$

By using the bound $w \leq (F + E + D)/3$ on the maximum number of holes we get:

$$\frac{5F^{\mathcal{P}} + 3E^{\mathcal{P}} + 3D^{\mathcal{P}} + 2C^{\mathcal{P}} + k - w}{(C^{\mathcal{P}} + D^{\mathcal{P}} + E^{\mathcal{P}} + 2F^{\mathcal{P}})/2 + k - w} =$$

$$\frac{10F^{\mathcal{P}} + 6E^{\mathcal{P}} + 6D^{\mathcal{P}} + 4C^{\mathcal{P}} + 2k - 2w}{2F^{\mathcal{P}} + E^{\mathcal{P}} + D^{\mathcal{P}} + C^{\mathcal{P}} + 2k - 2w} \leq$$

$$\frac{10F^{\mathcal{P}} + 6E^{\mathcal{P}} + 6D^{\mathcal{P}} + 4C^{\mathcal{P}} + 2k - 2(F^{\mathcal{P}} + E^{\mathcal{P}} + D^{\mathcal{P}})/3}{2F^{\mathcal{P}} + E^{\mathcal{P}} + D^{\mathcal{P}} + C^{\mathcal{P}} + 2k - 2(F^{\mathcal{P}} + E^{\mathcal{P}} + D^{\mathcal{P}})/3} =$$

$$\frac{28F^{\mathcal{P}} + 16E^{\mathcal{P}} + 16D^{\mathcal{P}} + 12C^{\mathcal{P}} + 6k}{4F^{\mathcal{P}} + E^{\mathcal{P}} + D^{\mathcal{P}} + 3C^{\mathcal{P}} + 6k} <$$

$$\frac{(28F^{\mathcal{P}} + 16E^{\mathcal{P}} + 16D^{\mathcal{P}} + 12C^{\mathcal{P}} + 6k) + (36F^{\mathcal{P}} + 36C^{\mathcal{P}} + 90k)}{4F^{\mathcal{P}} + E^{\mathcal{P}} + D^{\mathcal{P}} + 3C^{\mathcal{P}} + 6k} = 16.$$

Concerning the execution time, the algorithm performs $O(n)$ cuts. Each cut can be performed in $O(\log n)$-time by using the ray-shooting algorithm given in [1].   □

It is worth to note that Theorem 4 is based on Algorithm 1, and now this algorithm performs, as first step, a non-optimal decomposition of $\mathcal{P}$ into convex octilinear components. Unfortunately, Theorem 2 implies that such a convex decomposition cannot be performed efficiently.

7   Conclusion and Future Work

In this work we have introduced the OPD-TR problem, which consists in finding the minimal decomposition of an octilinear polygon with holes into basic components (octilinear triangles and rectangles). This problem is relevant in the context of modern electronic CAD systems, when the Cavity Model is used to detect the generation and propagation of electromagnetic noise into multi-layer PCBs. As main results, we have shown that the OPD-TR problem is NP-hard and proposed a constant-factor approximation algorithm.

We are currently implementing under CGAL the obtained 16-approximation algorithm. The implementation will be tested by using, as input, a large dataset from a real multi-layer PCB consisting of $\approx$100,000 total vertices.[1] The application domain (i.e., electronic CAD) and the size of such a dataset reveal that,

---

[1] The PCB consists of 16 layers and its $\approx$13,000 polygons (i.e., cavities) have been extracted from a Cadence® Allegro® PCB designer project file. The polygons have been approximated into octilinear polygons by using the schematization algorithm proposed in [3]. Disregarding the polygons having the area below a given threshold, we get the final dataset of $\approx$1,000 octilinear polygons with $\approx$100,000 total vertices.

for designing decomposition algorithms, we have to take into consideration not only the theoretical approximation bound but also the execution time. We feel that the experimentation will show that the proposed algorithm may succeed in both the aspects.

As future work, we propose to study the complexity (and to design decomposition algorithms) for the OPD-TR problem restricted to the case of octilinear polygons without holes, and to design approximate algorithms for the OPD-C problem.

References

1. Chazelle, B., Guibas, L.J.: Visibility and intersection problems in plane geometry. Discrete Comput. Geom. **4**, 551–581 (1989)
2. Cheng, X., Du, D.-Z., Kim, J.-M., Ruan, L.: Optimal rectangular partitions. In: Du, D.-Z., Pardalos, P.M. (eds.) Handbook of Combinatorial Optimization, pp. 313–327. Springer, New York (2005)
3. Cicerone, S., Cermignani, M.: Fast and simple approach for polygon schematization. In: Murgante, B., Gervasi, O., Misra, S., Nedjah, N., Rocha, A.M.A.C., Taniar, D., Apduhan, B.O. (eds.) ICCSA 2012, Part I. LNCS, vol. 7333, pp. 267–279. Springer, Heidelberg (2012)
4. Cicerone, S., Orlandi, A., Archambeault, B., Connor, S., Fan, J., Drewniak, J.L.: Cavities' identification algorithm for power integrity analysis of complex boards. In: 20th International Zurich Symposium on Electromagnetic Compatibility (EMC-Zurich 2009), pp. 253–256. IEEE Press (2009)
5. Ding-Zhu, D., Zhang, Y.: On heuristics for minimum length rectilinear partitions. Algorithmica **5**(1), 111–128 (1990)
6. Durocher, S., Mehrabi, S.: Computing partitions of rectilinear polygons with minimum stabbing number. In: Gudmundsson, J., Mestre, J., Viglas, T. (eds.) COCOON 2012. LNCS, vol. 7434, pp. 228–239. Springer, Heidelberg (2012)
7. Eppstein, D.: Graph-theoretic solutions to computational geometry problems. In: Paul, C., Habib, M. (eds.) WG 2009. LNCS, vol. 5911, pp. 1–16. Springer, Heidelberg (2010)
8. Ferrari, L.A., Sankar, P.V., Sklansky, J.: Minimal rectangular partitions of digitized blobs. Comput. Vis. Graph. Image Process. **28**(1), 58–71 (1984)
9. Imai, H., Asano, T.: Efficient algorithms for geometric graph search problems. SIAM J. Comput. **15**(2), 478–494 (1986)
10. Keil, J.M.: Polygon decomposition (Chap. 5). In: Sack, J.R., Urrutia, J. (eds.) Handbook on Computational Geometry, pp. 491–518. Elsevier Science, Amsterdam (2000)
11. Lei, C.T., Techentin, R.W., Gilbert, B.K.: High-frequency characterization of power/ground plane structures. IEEE Trans. Microw. Theory Tech. **47**, 562–569 (1999)
12. Lingas, A., Soltan, V.: Minimum convex partition of a polygon with holes by cuts in given directions. Theory Comput. Syst. **31**, 507–538 (1998)
13. Lipski, W.: An $O(n \log n)$ manhattan path algorithm. Inf. Process Lett. **19**(2), 99–102 (1984)
14. Lipsky, W., Lodi, E., Luccio, F., Mugnai, C., Pagli, L.: On two dimensional data organization II. Fundamenta Informaticae **2**, 245–260 (1979)

15. Na, N., Choi, J., Chun, S., Swaminatham, M., Srinivasan, J.: Modeling and transient Simulation of planes in electronic packages. IEEE Trans. Adv. Packag. **23**(3), 340–352 (2000)
16. Ohtsuki, T.: Minimum dissection of rectilinear regions. In: IEEE International Symposium on Circuits and Systems (1982)
17. Swaminathan, M., Joungho, K., Novak, I., Libous, J.P.: Power distribution networks for system-on-package: status and challenges. IEEE Trans. Adv. Packag. **27**(2), 286–300 (2004)

On Wrapping Spheres and Cubes
with Rectangular Paper

Alex Cole[1]([✉]), Erik D. Demaine[1], and Eli Fox-Epstein[2]

[1] MIT, Cambridge, MA, USA
alexcole@csail.mit.edu, edemaine@mit.edu
[2] Brown University, Providence, RI, USA
ef@cs.brown.edu

Abstract. What is the largest cube or sphere that a given rectangular piece of paper can wrap? This natural problem, which has plagued gift-wrappers everywhere, remains very much unsolved. Here we introduce new upper and lower bounds and consolidate previous results. Though these bounds rarely match, our results significantly reduce the gap.

1 Introduction

The problem of minimizing the amount of paper necessary to wrap a given 3-dimensional surface arises naturally from the economics of any factory packaging physical items. We study two closely related cases of this problem: given an $x \times 1/x$ unit-area rectangle of paper, what is the largest possible cube or sphere it can wrap?

Informally, we consider wrappings that do not stretch, cut, or intersect the paper with itself. We allow multiple layers of paper in the folding, and unlike [3], do not differentiate between the front and back of the paper. As in [5], to formally capture what it means to wrap a surface with non-zero curvature, we define a *wrapping* (a.k.a. *folding*) to be a *noncrossing, contractive mapping* from a 2-dimensional rectangle of paper to a subset of Euclidean 3-space. A contractive function ensures no pair of points move apart under their image. This definition captures the "crinkling" that you observe when, for example, you physically wrap a billiard ball with a sheet of paper. The noncrossing condition is difficult to formalize (see [7]); intuitively, it prohibits wrappings that cause surfaces to strictly intersect.

We present a variety of novel techniques that improve upper and lower bounds for wrapping both spheres and cubes. Figure 1 graphically summarizes these results. A *sphere wrapping* (respectively, *cube wrapping*) is a wrapping whose image is a sphere (cube). We denote cubes of sidelength S as S-cubes and spheres of radii R as R-spheres. Throughout, assume that $0 < x \leq 1$ so that x is the smaller side of our $x \times 1/x$ paper.

Eli Fox-Epstein: Supported in part by NSF Grant CCF-0964037.

J. Akiyama et al. (Eds.): JCDCGG 2013, LNCS 8845, pp. 31–43, 2014.
DOI: 10.1007/978-3-319-13287-7_4

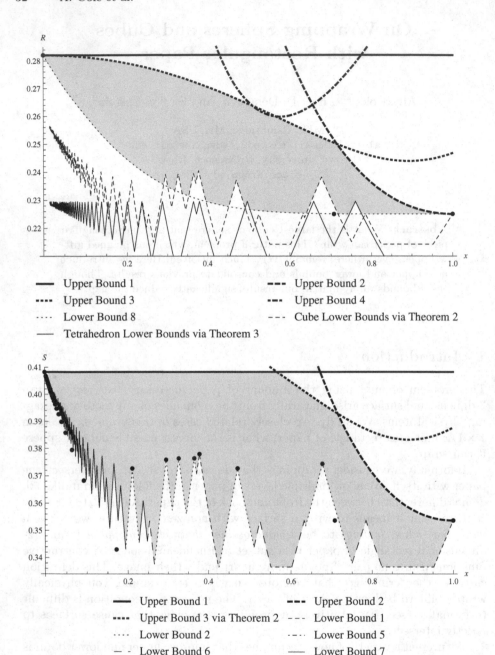

Fig. 1. A summary of the upper bounds (thick lines) and lower bounds (thin lines) on sphere (top) and cube (bottom) foldings. The horizontal axes indicate the smaller dimension of the unit-area paper. The vertical axes denote radii and side lengths, respectively. The shaded regions indicate the gaps between the bounds. See Sect. 6 for discussion.

2   Previous Results

2.1   Upper Bounds

Two techniques generated the previously known upper bounds on wrappings of spheres and cubes. The first is the surface area bound: the surface area of the image of a contractive mapping cannot exceed the surface area of the paper.

Upper Bound 1 (Folklore). *A unit-area rectangle of paper may wrap an S-sphere or an R-cube only if $S \leq 1/\sqrt{6}$ and $R \leq 1/(2\sqrt{\pi})$.*

Catalano-Johnson and Loeb [4] observe that every point on the S-cube has an antipodal point at least $2S$ away. Because wrappings are contractive, every point (particularly the center) on the original paper must also have another point that is $2S$ away, implying the paper's diagonal is at least $4S$. Demaine et al. [5] apply this argument to spheres.

Upper Bound 2 [4,5] . *An $x \times 1/x$ rectangle of paper may wrap an S-sphere or an R-cube only if $S \leq \sqrt{x^2 + x^{-2}}/4$ and $R \leq \sqrt{x^2 + x^{-2}}/(2\pi)$.*

The surface area bound becomes tight as x approaches 0 and the antipodal points bound is tight when $x = 1$. Between the endpoints, these bounds are likely far from optimal.

2.2   Lower Bounds

Numerous lower bounds for particular rectangles, some with unclear origins, exist in the form of physical foldings.

Lower Bound 1 ([4,9], Folklore). $1/\sqrt{7} \times \sqrt{7}$ *paper wraps a $1/\sqrt{7}$-cube, and 1×1 and $1/\sqrt{2} \times \sqrt{2}$ papers each wrap a $1/(2\sqrt{2})$-cube.*

Akiyama, Ooya, and Segawa [3] produce a series of six efficient "symmetric-skew" wrappings which spiral the paper around the cube.

Lower Bound 2 [3]. *$x \times 1/x$ paper wraps an S-cube for each (x, S) pair:*

$$\left(\sqrt{\tfrac{11}{24}}, \sqrt{\tfrac{37}{264}} \right), \left(\sqrt{\tfrac{2}{9}}, \sqrt{\tfrac{5}{36}} \right), \left(\sqrt{\tfrac{2}{15}}, \sqrt{\tfrac{17}{120}} \right),$$

$$\left(\sqrt{\tfrac{8}{75}}, \sqrt{\tfrac{17}{120}} \right), \left(\sqrt{\tfrac{2}{23}}, \sqrt{\tfrac{13}{92}} \right), \left(\sqrt{\tfrac{2}{45}}, \sqrt{\tfrac{5}{36}} \right).$$

Akiyama et al. [3] also invent a technique called *strip folding*, using extremely long, narrow rectangles to come arbitrarily close to the surface area bound.

Lower Bound 3 [3]. *A strip of paper with $x = 1/\sqrt{24n^2 + 12n - 2}$ can wrap a $2n/\sqrt{24n^2 + 12n - 2}$-cube for integers $n \geq 1$.*

Demaine et al. [6] revisit strip folding, showing that any polyhedron can be wrapped by strip folding.

Sphere wrappings are less extensively studied than cube wrappings.

Lower Bound 4 [5]. 1×1 and $1/\sqrt{2} \times \sqrt{2}$ *rectangles wrap a* $1/(\pi\sqrt{2})$-*sphere.*

Demaine et al. [5] also apply strip folding to spheres but do not provide an explicit construction.

3   Upper Bounds

The following two techniques create new upper bounds on sphere wrapping and provide a substantial improvement over the previous upper bounds, as illustrated in Fig. 1. Our general approach is to reduce the problem of bounding rectangular wrappings to simpler shapes like circles and stadiums.

3.1   Inscribed Stadiums on Spheres

As x approaches 1, Upper Bound 1 (surface area) becomes less effective, as it fails to account for necessary "crumpling" of the paper. This technique improves upon this by observing that a particular shape inscribed within paper must waste a certain amount of its surface area when mapped onto a sphere.

An $x \times y$ *stadium* is the Minkowski sum of a length-x line segment (called the *major path*) with a diameter-y disk. Refer to Fig. 2.

Proposition 1. *Given an* $x \times y$ *stadium* S *with major path* P *mapped onto a sphere by some contractive function* f, *let* X *be the points on the sphere within surface distance* $y/2$ *from* $f(P)$. *Then* $f(S) \subset X$.

Proof. On the flat paper, all of the points in S are within $y/2$ of the major path P. Because f is contractive, all of these distances can only decrease when S is mapped onto the sphere.   □

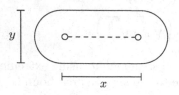

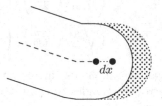

Fig. 2. $x \times y$ stadium, dashed major path.

Fig. 3. Extension of a stadium by dx.

Proposition 2. *An $x \times y$ stadium of flat paper mapped onto an R-sphere may occupy no more surface area than*

$$A(x, y) = 2R \left(\pi R - \pi R \cos \frac{y}{2R} + x \sin \frac{y}{2R} \right).$$

Proof. To bound $A(x, y)$ we will first establish $A(0, y)$ and then bound the derivative dA/dx. This will allow us to bound the areas of all mapped stadiums.

First, consider an $0 \times y$ stadium: a radius-$y/2$ disk. By definition, the disk must fall within $y/2$ of its center on the sphere. A radius-$y/2$ spherical cap has area $2\pi R^2 (1 - \cos \frac{y}{2R})$, proving the claim for $A(0, y)$.

Now consider an $x \times y$ stadium S with major path P. Let f be a contractive map to the sphere and A be the area of points within distance $y/2$ of $f(P)$ on the sphere. From Proposition 1, it suffices to bound A. Extend S by some length dx (see Fig. 3). For sufficiently small dx, the extension of $f(P)$ runs along a geodesic. Call the added area dA. In Fig. 3, this is the dotted region.

Now let $\theta = dx/R$. This is the central angle corresponding to a geodesic of length dx on the sphere. Extending P by dx will affect the latitudes within $y/2$ of our geodesic. Each latitude can be extended by at most an angle of θ. Let r_a be the radius of a circle of latitude at a spherical distance a from the equator. It is well-known that $r_a = R \cos \frac{a}{R}$. This yields:

$$dA \leq \int_{-y/2}^{y/2} \theta r_a \, da - \int_{-y/2}^{y/2} \frac{dx}{R} R \cos \frac{a}{R} \, da = 2R \sin \frac{y}{2R} dx$$

so $dA/dx \leq 2R \sin y/(2R)$. For a stadium of length x, the area on the sphere is bounded by $A(x, y)$ as desired. □

A $(1/x - x) \times x$ stadium can be inscribed within any $x \times 1/x$ paper rectangle. By Proposition 2, this stadium only occupies $A(1/x - x, x)$ area on the sphere. The remaining paper only has an area of $x^2 - \pi x^2/4$.

Upper Bound 3. *$x \times 1/x$ paper can wrap an R-sphere only if*

$$4\pi R^2 \leq x^2 - \pi x^2/4 + A(1/x - x, x).$$

3.2 n Circumscribed Circles on Spheres

Cutting the paper or adding more material can only increase the ability of the paper to wrap a sphere. With this as inspiration, we transform the paper into n congruent disks, and then relate upper bounds on spherical cap coverings back to rectangular sphere wrappings.

Proposition 3. *If an R-sphere can be wrapped by an $x \times 1/x$ paper, then it can also be wrapped by n congruent disks of diameter $\sqrt{x^2 + (nx)^{-2}}$.*

Proof. Consider a arbitrary wrapping from an $x \times 1/x$ paper to an R-sphere. Partition the flat paper into n small $x \times 1/(nx)$ rectangles. Circumscribe each $x \times 1/(nx)$ rectangle to get n disks of diameter $\sqrt{x^2 + (nx)^{-2}}$. These disks can contractively map into the original paper. □

Covering a sphere with n spherical caps is a well-studied problem (see e.g. [8,11]) and for many values of n, bounds exist on how large the diameter d must be to admit a covering of a sphere of radius R. For $n = 1$, a disk of diameter d can wrap an R-sphere only if $d \geq 2\pi R$. This yields $\sqrt{x^2 + x^{-2}} \geq 2\pi R$, which is exactly Upper Bound 2! This generalization of the antipodal points bound is most useful when n is 1 or 3, where coverings are necessarily very wasteful.

Upper Bound 4. *An $x \times 1/x$ rectangle may wrap an R-sphere only if $R \leq \sqrt{x^2 + (3x)^{-2}}/\pi$.*

Proof. Three diameter-d disks can wrap an R-sphere only if $d \geq \pi R$ (see s Table 2 of [11]). Composing with the contrapositive of Proposition 3 yields $R \leq \sqrt{x^2 + (3x)^{-2}}/\pi$.   □

4   Lower Bounds

4.1   Rescaling Lower Bounds on Cubes

Most lower bounds on wrapping take the form of a construction for a specific x. Here we present a method to rescale particular foldings to produce a continuous set of lower bounds.

Theorem 1. *If $x \times 1/x$ paper wraps an S-cube, then there exists a folding of an $x' \times 1/x'$ rectangle into an $f(x')$-cube where $f(x') = S\min\{x'/x, x/x'\}$.*

Proof. Suppose $x' < x$. Uniformly scaling an $x \times 1/x$ rectangle by a factor of x'/x, yields an $x' \times x'/x^2$ rectangle, which wraps an Sx'/x-cube. An $x' \times 1/x'$ rectangle contracts to an $x' \times x'/x^2$ rectangle. A corresponding argument can be made for $x' > x$.   □

4.2   Rectangle Conversions on Cubes

The rectangle-to-rectangle hinged dissection gadget of [1] inspires a technique to transform wrappings of a particular aspect ratio into general wrappings without any loss of efficiency.

Lower Bound 5. *Any unit-area rectangle wraps a $1/\sqrt{6 + 2\sqrt{2}}$-cube.*

Proof. The crease pattern in the top-left of Fig. 4 shows a valid wrapping f of a $1/\sqrt{6 + 2\sqrt{2}}$-cube from an $x \times 1/x$ rectangle (fold each horizontal or vertical crease to a right angle) with these special properties:

1. Left and right edges of the paper map to same segment.
2. Left and right halves of the top edge map to the same segment.
3. The bottom edge maps to a point.

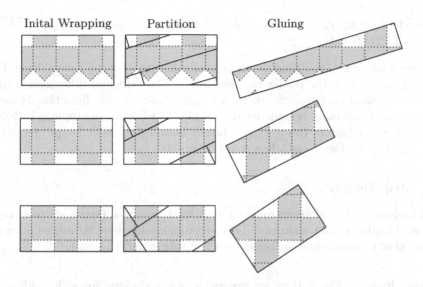

Initial Wrapping   Partition   Gluing

Fig. 4. The transformations used in Lower Bounds 5 (top) and 6 (middle and bottom).

Wrapping f can be transformed into a wrapping from a different aspect ratio rectangle as follows. Partition an $x \times 1/x$ rectangle into some set of pieces P. Notice that one can still wrap a $1/\sqrt{6+2\sqrt{2}}$-cube with P by applying f to each part.

To create a new rectangular wrapping, we will "glue" P back together by identifying edges of elements of P such that they form an $x' \times 1/x'$ rectangle. To ensure this gluing produces a valid wrapping, identified points must map to the same point under f. Then f applied to each piece of P results in a contractive mapping f' from $x' \times 1/x'$ to the $1/\sqrt{6+2\sqrt{2}}$-cube.

To partition an $x \times 1/x$ rectangle and glue it into an $x' \times 1/x'$ rectangle, Montucla's dissection from [10] suffices. This is visualized in step 2 of Fig. 4.

To glue the parts back together:

1. Identify the original left edge and the original right edge.
2. Identify the left half of the original top edge and the right half.
3. Identify the two small parts of the bottom edge with the large part.

Figure 4 illustrates this transformation. The first and second identifications correspond directly to special properties 1 and 2. The last one is valid because the entire bottom edge is mapped to a single point.

Varying the angle of the diagonal cut in the dissection creates rectangles of any aspect ratio.   □

The efficiency of Lower Bound 5 can be increased by starting with a different wrapping. This causes the cuts to become more constrained, resulting in a discrete set of wrappings. This technique is similar to the tetrahedral wrappings in [2].

Lower Bound 6. *For any integer $n \geq 2$, a rectangle with $x = 2\sqrt{2}/\sqrt{n^2 + 4}$ wraps a $1/(2\sqrt{2})$-cube.*

The proof of this proposition is almost identical to that of Lower Bound 5. Figure 4 should give the reader the core ideas. The principal difference is that the bottom edge no longer maps to a single point, so the dissection is more constrained. Combining bounds on the middle and bottom wrappings in Fig. 4 yields our lower bound. Interestingly, Lower Bound 6 for $n = 2$ reproduces the square folding by Catalano-Johnson and Loeb [4].

4.3   Strip Folding

Strip folding is a technique introduced in [3] that weaves a narrow strip of paper back and forth to cover a surface. This section sketches new strategies for strip folding that produce superior bounds on the sphere and the cube.

Cubes. Refer to Fig. 5. Here we present a new technique for strip folding on the cube that is more efficient than that presented in [3]. The general strategy consists of 3 parts resembling an algorithm more than a function:

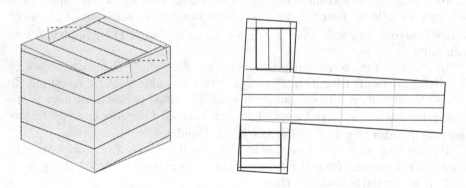

Fig. 5. Strip wrapping a cube. 3D diagram (left) and edge unfolding (right).

– Spiral around the 4 vertical faces of the cube (sides).
– Fold the excess over onto top and bottom faces.
– Try two different methods of doubling back and forth using turn gadgets (as seen in [3]) to cover the rest of the top and bottom faces.

We parameterize in terms of n, the number of times the top of the strip switches faces while covering the sides. For ease of computation, we require integral n.

The excess folded onto the top and bottom leaves a $w \times h$ rectangle uncovered on each, where $w = \frac{(n-5)S}{\sqrt{1+n^2}}$ and $h = \frac{(n-3)S}{\sqrt{1+n^2}}$. In Fig. 5, the bold lines indicate these $w \times h$ rectangles.

Proposition 4. *A* $w \times h$ *rectangle can be covered by an* $x \times f(w, h, x)$ *rectangle of paper when one end of the paper starts outside the corner of the rectangle along the side of length* h *where*

$$f(w, h, x) = \min\left(2x\lceil h/x \rceil + w\lceil h/x \rceil - x, 2x\lceil w/x \rceil + h\lceil w/x \rceil\right).$$

Both parts of the minimum have the strip move straight in one direction until it would overlap some paper, then turning at a right angle twice to double back. Both terms start with an initial turn (or two) to enter the rectangle and orient to run parallel to a side. The second term is improved to $w + \left\lceil \frac{w}{x} \right\rceil (h + x)$ with a slightly more complicated construction.

After fixing n, we need two equations to solve for the width x and sidelength S. Combining the three lengths in the construction:

$$\underbrace{1/x}_{\text{total length}} = \underbrace{S\sqrt{1 + n^2} + \sqrt{16S^2 - x^2}}_{\text{length to cover sides}} + \underbrace{2f\left(\frac{(n-5)S}{\sqrt{1+n^2}}, \frac{(n-3)S}{\sqrt{1+n^2}}, x\right)}_{\text{extra length for top and bottom}}. \quad (1)$$

Unrolling the spiraling portion of the strip and using similar triangles yields

$$x/S = 4/\sqrt{1 + n^2}. \quad (2)$$

Lower Bound 7. *For any integer* $n \geq 5$, $x \times 1/x$ *paper wraps an* S-*cube where* S *satisfies Eqs. (1) and (2).*

When $n = 5$ and $n = 7$, we recover two of the foldings from [3].

Spheres. Any point on a sphere of fixed radius can be described by the two angles of spherical coordinates: θ (polar angle) and ϕ (azimuthal angle). The underlying strategy is to spiral with constant slope much in the way the 4 sides of the cube were wrapped. Figure 6 shows how the strip wraps a sphere by

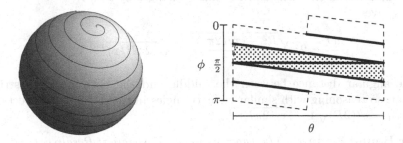

Fig. 6. Strip wrapping a sphere. Bold lines are mapped without any contraction.

maintaining constant $d\theta/d\phi$. We focus on only the top hemisphere as the bottom follows by symmetry.

Start with an *initialization rectangle* of width x with the diagonal $2\pi R$ wrapped exactly around an equator. This rectangle is indicated by the dotted region in Fig. 6. The top edge forms a line segment with $\frac{d\theta}{d\phi}$ held constant.

Now extend the initialization rectangle, spiraling up the sphere continuing to hold $\frac{d\theta}{d\phi}$ constant. Terminate when the center of the ends reaches the poles.

A "cut and paste" argument, similar to that used for Lower Bound 5, rearranges the ends of the strip to ensure the poles are covered. This is visualized in Fig. 6 by moving the paper above $\phi = 0$ into the gap at the top and doing the same for $\phi = \pi$.

The length of the strip, $1/x$, is the length of the initialization rectangle plus the amount needed to spiral in each hemisphere: $1/x = L_{\text{init}} + 2L_{\text{spiral}}$. In calculating the lengths, care must be taken to only look at fully stretched paths because the rest of the strip is being contracted. These paths are bold in Fig. 6.

The initialization rectangle has diagonal $2\pi R$ and height x, so its length is given by $L_{\text{init}} = \sqrt{(2\pi R)^2 - x^2}$.

Consider the upper hemisphere. By construction, in the upper hemisphere, the bottom of the strip incurs no contraction. Using the spherical arc length formula:

$$dl^2 = (R\sin(\phi)d\theta)^2 + (Rd\phi)^2 + dR^2 = Rd\phi\sqrt{\sin^2(\phi)\left(\frac{d\theta}{d\phi}\right)^2 + 1}$$

Integrating over ϕ gives the length:

$$L_{\text{spiral}} = \int dl = R\int_{\phi_0}^{\frac{\pi}{2}} \left(\sqrt{\sin^2(\phi)\left(\frac{d\theta}{d\phi}\right)^2 + 1}\right) d\phi$$

where ϕ_0 is the value of ϕ such that when the bottom of the strip is at an angle ϕ_0, the middle of the strip reaches the pole.

The slope $d\theta/d\phi$ is constant and thus equal to the ratio of the total change in θ to the total change in ϕ over the initialization rectangle. θ ranges from 0 to 2π. Similar triangles demonstrate that ϕ changes by $x/\sqrt{R^2 - (x/(2\pi))^2}$, so dividing gives

$$\frac{d\theta}{d\phi} = 2\pi\frac{\sqrt{R^2 - (x/(2\pi))^2}}{x} = \sqrt{(2\pi R/x)^2 - 1}.$$

The angular distance between the middle and the bottom of the strip is constant, so reasoning with similar right triangles in the initialization rectangle yields $\phi_0 = x/(2R)\sqrt{1 - (x/(2\pi R))^2}$.

Lower Bound 8. *An $x \times 1/x$ paper wraps an R-sphere if R satisfies*

$$\frac{1}{x} = \underbrace{\sqrt{(2\pi R)^2 - x^2}}_{init} + \underbrace{2R\int_{\frac{x}{2R}\sqrt{1-\left(\frac{x}{2\pi R}\right)^2}}^{\frac{\pi}{2}} \left(\sqrt{\sin^2(\phi)\left((2\pi R/x)^2 - 1\right) + 1}\right) d\phi}_{each\ hemisphere}$$

This bound is the first explicit sphere strip folding. It becomes optimally efficient as x tends to 0. In addition, because of how it handles the ends of the strip, it provides a powerful lower bound for larger values of x. When $x = 1$ we recover the optimal lower bound from [5].

5   Relating Cubes and Spheres

Given contractive mappings $f : A \to B$ and $g : B \to C$, $g \circ f$ constitutes a valid contractive mapping from A to C. With this as inspiration, we present mappings between spheres and cubes, allowing upper and lower bounds for one to be translated to the other. Upper Bound 3 for inscribed stadiums translates particularly well.

Theorem 2. *S-cubes can be contractively mapped to $(2S/\pi)$-spheres.*

Proof. Let f be our contractive mapping. Consider the Voronoi regions on a sphere induced by the six x-, y-, and z-extremal points. f will contractively map each face of the cube into one of these regions.

Fig. 7. One face of a cube with the area used by g drawn in.

It suffices to examine one face F and the corresponding sixth of a sphere F'. Refer to Fig. 7. Let the center of F be $(0,0)$ and the center of F' be $(0,0,R)$. Let $g : F' \to F$ be the map that sends a point with spherical coordinates $x = (R, \theta, \phi)$ to the polar point $g(x) = (R\phi, \theta)$ on the paper.

Now we will show g is expansive by looking at an infinitesimal neighborhood of an arbitrary x. Let $x = (R, \theta, \phi)$ and $\tilde{x} = (R, \theta + d\theta, \phi + d\phi)$. Now let $dl_1 = \|x - \tilde{x}\|$ and $dl_2 = \|g(x) - g(\tilde{x})\|$. These are known as line elements. It is well-known that the sphere metric yields $dl_1^2 = (R\sin\phi d\theta)^2 + (Rd\phi)^2$. Doing the same about $g(x)$ with the metric on the paper gives us $dl_2^2 = (Rd\phi)^2 + (R\phi d\theta)^2$. Because $\sin^2\phi \leq \phi^2$, $dl_1 \leq dl_2$. These distances can be integrated into arclengths to show that all pairwise distances on the sphere are less than their images on the paper. Thus g is expansive, so $f = g^{-1}$ is contractive. The image of g is just a subset of F, but we can extend the domain of f to all of F by mapping the unused region to the boundary of F'.

The map f sends the line going through the centers of four faces of the cube to an equator of the sphere without any contraction. If S is the sidelength of the cube, then the resulting sphere will have radius $R = 2S/\pi$. This also shows f is optimal: no contractive mapping can produce larger spheres from a cube. $\square$

Theorem 3. *S-tetrahedra contractively map to $S/(2\sqrt{3}\arccos\frac{1}{\sqrt{3}})$-spheres.*

Theorem 3 is proved similarly to Theorem 2. Composing the tetrahedral wrappings in [2] with Theorem 3 improves the sphere lower bounds in some regions, as shown in Fig. 1.

6   Conclusions

Figure 1 gives a complete picture of the upper and lower bounds on spheres and cubes, respectively. The horizontal axes denote the short paper dimension x. The vertical axes give radius R (for the sphere) and the sidelength S (for the cube). The shaded region is the area where the largest radius/sidelength could lie. To simplify presentation, lower bounds on cube foldings are only displayed when they are the best known. Previous lower bounds (Lower Bounds 1–4) are plotted as black dots.

Using Theorem 1, discrete constructions for cube lower bounds are transformed into a continuum. One surprise here is that the $1/\sqrt{2} \times \sqrt{2}$ wrapping of the $1/(2\sqrt{2})$-cube is less efficient than a rescaling of a construction from Lower Bound 2. Other results in Sect. 4 provided significant improvements over previous known bounds across a variety of aspect ratios.

The two new spherical upper bounds from Sect. 3 greatly improve upon previous bounds, especially for intermediate values of x. Upper Bound 3 (inscribed stadiums), in particular, is such an improvement that it transfers to the cube using contractive mappings, creating the first new cube upper bound since 2001. Upper Bound 4 is generally weaker but still provides an improvement for some aspect ratios. Quite surprisingly, composing Upper Bound 4 with Theorem 2 gives an upper bound tangent to Upper Bound 1 (surface area). Finally, the contractive mappings translate cube and tetrahedron wrappings to the sphere, elevating the lower bounds.

Acknowledgments. This research began in an open problem session and final project for MIT class 6.849: Geometric Folding Algorithms in Fall 2012. Thanks to Stephen Face for fruitful discussion and to Zachary Abel and Martin Demaine for their assistance with references.

References

1. Abbott, T.G., Abel, Z., Charlton, D., Demaine, E.D., Demaine, M.L., Kominers, S.: Hinged dissections exist. Discrete Comput. Geom. **47**(1), 150–186 (2012)
2. Akiyama, J., Hirata, K., Kobayashi, M., Nakamura, G.: Convex developments of a regular tetrahedron. Comput. Geom. Theory Appl. **34**(1), 2–10 (2006)
3. Akiyama, J., Ooya, T., Segawa, Y.: Wrapping a cube. Teach. Math. Appl. **16**(3), 95–100 (1997)
4. Catalano-Johnson, M.L., Loeb, D., Beebee, J.: Problem 10716: A cubical gift. Am. Math. Monthly **108**(1), 81–82 (2001)

5. Demaine, E.D., Demaine, M.L., Iacono, J., Langerman, S.: Wrapping spheres with flat paper. Comput. Geom. Theory Appl. **42**(8), 748–757 (2009)
6. Demaine, E.D., Demaine, M.L., Mitchell, J.S.B.: Folding flat silhouettes and wrapping polyhedral packages: New results in computational origami. Comput. Geom. Theory Appl. **16**(1), 3–21 (2000)
7. Demaine, E.D., O'Rourke, J.: Geometric Folding Algorithms: Linkages, Origami, Polyhedra, pp. 179–182. Cambridge University Press, New York (2007)
8. Fowler, P.W., Tarnai, T.: Transition from circle packing to covering on a sphere: the odd case of 13 circles. Proc. Royal Soc. London. Ser. A: Math. Phys. Eng. Sci. **455**, 4131–4143 (1999)
9. Gardner, M.: New Mathematical Diversions, Chapter 5: Paper Cutting, pp. 58–69. Simon and Schuster, New York (1966)
10. Ozanam, J.: Récréations Mathématiques et Physiques, pp. 297–302. C.A. Jombert (1790)
11. Tarnai, T., Gáspár, Z.: Covering a sphere by equal circles, and the rigidity of its graph. Math. Proc. Camb. Philos. Soc. **110**, 71–89, 7 (1991)

On Polygonal Paths with Bounded Discrete-Curvature: The Inflection-Free Case

Sylvester Eriksson-Bique[1], David Kirkpatrick[2]([✉]), and Valentin Polishchuk[3]

[1] Courant Institute, New York, USA
seb522@nyu.edu
[2] University of British Columbia, Vancouver, Canada
kirk@cs.ubc.ca
[3] University of Helsinki, Helsinki, Finland
polishch@cs.helsinki.fi

Abstract. A shortest path joining two specified endpoint configurations that is constrained to have mean curvature at most ς on every non-zero length sub-path is called a ς-*geodesic*. A seminal result in non-holonomic motion planning is that (in the absence of obstacles) a 1-geodesic consists of either (i) a (unit-radius) circular arc followed by a straight segment followed by another circular arc, or (ii) a sequence of three circular arcs the second of which has length at least π [Dubins, 1957]. Dubins' original proof uses advanced calculus; Dubins' result was subsequently rederived using control theory techniques [Sussmann and Tang, 1991], [Boissonnat, Cérézo, and Leblond, 1994], and generalized to include reversals [Reeds and Shepp, 1990].

We introduce and study a discrete analogue of curvature-constrained motion. Our overall goal is to show that shortest polygonal paths of bounded "discrete-curvature" have the same structure as ς-geodesics, and to show that properties of ς-geodesics follow from their discrete analogues as a limiting case, thereby providing a new, and arguably simpler, "discrete" proof of the Dubins characterization. Our focus, in this paper, is on paths that have non-negative mean curvature everywhere; in other words, paths that are free of inflections, points where the curvature changes sign. Such paths are interesting in their own right (for example, they include an additional form, not part of Dubins' characterization), but they also provide a slightly simpler context to introduce all of the tools that will be needed to address the general case in which inflections are permitted.

1 Introduction

Curvature-constrained paths are a fundamental tool in planning motion with bounded turning radius. Paths that are *smooth* (continuously differentiable) have the advantage that they may look more appealing and realistic than *polygonal* (piecewise-linear) paths. Nevertheless, polygonal paths are a much more common model in geometry, exactly because of their discrete nature, and for this same reason they have the potential of providing simpler and more intuitive proofs

© Springer International Publishing Switzerland 2014
J. Akiyama et al. (Eds.): JCDCGG 2013, LNCS 8845, pp. 44–64, 2014.
DOI: 10.1007/978-3-319-13287-7_5

of properties of their smooth counterparts. Furthermore, from an applications perspective, polygonal paths are more natural to plan, describe and follow. For instance, in air traffic management—one of our motivating applications—an aircraft flight plan (a list of "waypoints") is represented on the strategic level by a polygonal path whose vertices are the GPS waypoints. The actual smoothly turning trajectory at a waypoint is decided by the pilot on the tactical level when executing the turn (see [25] for more on curvature-constrained route planning in air transport). We are thus motivated to formulate a *discretized* model of curvature-constrained motion.

Smooth paths of bounded curvature. In studying smooth paths of bounded curvature, L. E. Dubins [15] observed that if one restricts attention to paths whose curvature is defined at every point then there are situations in which shortest paths do not exist. On the other hand, if one only requires that paths are everywhere differentiable (that is their slope is well-defined at every point) then their *mean* curvature is well-defined on every non-zero length sub-path. Thus, Dubins chose to define a path to have bounded curvature if its mean curvature is bounded everywhere. Specifically, let γ be a smooth path, parameterized by its arclength. For any t in the domain of γ, let $\gamma'(t)$ denote the derivative of γ at t – the unit vector tangent to γ at t. The path γ is said to have mean curvature at most one if $\angle st \leq s - t$, for any $t < s < t + \pi$, where $\angle st$ denotes the angle between the directions of $\gamma'(s)$ and $\gamma'(t)$. (In other words, γ', viewed as mapping from the domain of γ to the unit circle, is 1-Lipschitz.) Furthermore, there is no loss of generality in restricting attention to the case where the mean curvature bound is one, since the general case can be reduced to the unit case by suitable scaling. Accordingly, we hereafter use the term "curvature-constrained" as a shorthand for "has mean curvature bounded by one", and we refer to paths with this property as *cc-paths*.

Dubins' characterization, a seminal result in curvature-constrained motion planning, states that, in the absence of obstacles, shortest curvature-constrained paths in the plane, are one of two types: (i) a circular arc followed by a line segment followed by another arc, or (ii) a sequence of three circular arcs, the second of which has length at least π.

Discrete circular arcs. While several possibilities suggest themselves as ways to formulate a discrete analogue of unit-bounded curvature[1], it seems that all such formulations are based on a natural notion of discrete circular arcs. Let $0 < \theta \leq \pi/2$ be a given angle. We say that a polygonal chain forms a θ-*discrete circular arc* (or simply a discrete circular arc if θ is understood) if (i) its vertices belong, in sequence, to a common circle of radius one, and (ii) successive edges have length at most $d_\theta = 2\sin\frac{\theta}{2}$ (that is they subtend a circular arc of length at most θ). Any portion of regular polygon with $k \geq 3$ sides, inscribed in a unit circle, provides a prototypical $(2\pi/k)$-discrete circular arc.

---

[1] In an earlier draft [18], the authors proposed an alternative definition which had some deficiencies that are resolved by the definition used in this paper.

Polygonal paths of bounded discrete-curvature. Discrete circular arcs not only satisfy our intuitive notion of polygonal path with bounded discrete-curvature but, like their smooth counterparts, they seem to capture the extreme case. It is interesting to ask what properties a path P with bounded discrete-curvature should have in general. To ensure that such a path does not turn "too sharply" it seems natural to require that, like discrete circular arcs, P should turn by at most θ at each of its interior vertices. However, such a restriction alone does not guarantee that P will serve as a bona fife discrete analogue of a bounded-curvature path: many short successive edges of P, each

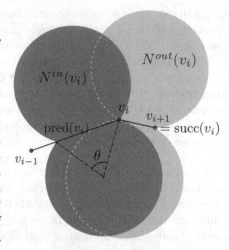

Fig. 1. Local conditions for θ-discrete curvature-constrained paths.

turning only slightly, can simulate a sharp turn. Taking further inspiration from discrete circular arcs, that permit short edges with correspondingly gentle turns, we define:

Definition 1. *A polygonal path $\langle v_1, v_2, \ldots, v_n \rangle$ has θ-discrete-curvature (or just discrete-curvature, if θ is understood) at most one if, for $1 < i < n$, the turn at vi, the difference between the angles of the rays $v_{i-1}v_i$ and v_iv_{i+1}, is no more than $\sin^{-1}\left(\frac{\min\{d_\theta, |v_{i-1}v_i|\}}{2}\right) + \sin^{-1}\left(\frac{\min\{d_\theta, |v_iv_{i+1}|\}}{2}\right)$.*

Remark 1. As with its smooth counterpart, we hereafter use the term "θ-discrete-curvature-constrained" (frequently abbreviated as "θ-dcc") to mean "has θ-discrete-curvature bounded by one". It is easy to confirm that (i) any discrete circular arc is a dcc-path, and (ii) for every vertex v_i on a dcc-path, the point $\text{succ}(v_i)$ at distance $\min\{d_\theta, |v_iv_{i+1}|\}$ from v_i along the edge v_iv_{i+1}, does not lie in the interior of the region $N^{in}(v_i)$ formed by the union of the unit circles passing through v_i and $\text{pred}(v_i)$, the point at distance $\min\{d_\theta, |v_{i-1}v_i|\}$ from v_i along the edge $v_{i-1}v_i$. We will refer to $N^{in}(v_i)$ (respectively, $N^{out}(v_i)$), the region formed by union of the unit circles passing through v_i and $\text{succ}(v_i)$ as the *in-neighbourhood* (respectively, *out-neighbourhood*) of v_i (see Fig. 1). Note that, if $|v_iv_{i+1}| \leq d_\theta$ then $N^{out}(v_i) = N^{in}(v_{i+1})$.

Relating smooth and discrete curvature-constrained paths. Let γ be a smooth path, parameterized by its arclength. We say that a polygonal path $\hat{\gamma} = \langle v_1, v_2, \ldots, v_n \rangle$ is a θ-*discretization* of γ if (i) $v_i = \gamma(t_i)$, for $1 \leq i \leq n$, where (ii) $t_1 = 0$, $t_n = |\gamma|$, and $0 \leq t_{i+1} - t_i \leq \theta$, for $1 \leq i < n$.

By definition, a θ-discrete circular arc is a θ-discretization of an arc of a (smooth) unit circle. In fact (cf. Theorem 1 below) *every θ-discretization $\hat{\gamma} = \langle \gamma(t_1), \gamma(t_2), \ldots, \gamma(t_n) \rangle$ of every cc-path γ forms a θ-dcc-path.*

Lemma 1. *For any r and s, $r < s < r + \pi$, in the domain of γ, $|\gamma(s) - \gamma(r)| \geq 2 \sin \frac{s-r}{2}$.*

Proof. Assume, without loss of generality, that $r = 0$ and that $\gamma(r)$ is at the origin O (i.e. $\gamma(0) = (0,0)$); the lemma is then equivalent to $|\gamma(s)| \geq 2 \sin \frac{s}{2}$ (Fig. 2). We prove this by lower-bounding the derivative of $|\gamma(s)|^2$:

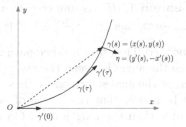

Fig. 2. $|O\gamma(s)| \geq 2 \sin \frac{s}{2}$, the slope of $O\gamma(s)$ is at most $\tan \frac{s}{2}$.

$$(|\gamma(s)|^2)' = 2\gamma(s) \cdot \gamma'(s) = 2\int_0^s \gamma'(\tau) \cdot \gamma'(s)\, d\tau = 2\int_0^s \cos \angle s\tau \, d\tau \geq 2\int_0^s \cos(s - \tau)\, d\tau = 2\sin s$$

Hence $|\gamma(s)|^2 \geq 2(1 - \cos s) = 4\sin^2 \frac{s}{2}$. $\qquad\square$

Corollary 1. *For all i, $1 \leq i < n$, $|\gamma(t_{i+1}) - \gamma(t_i)| \geq |t_{i+1} - t_i| \frac{\sin(\theta/2)}{\theta/2}$.*

Proof. It suffices to observe that, since $0 \leq t_{i+1} - t_i \leq \theta$, $\frac{\sin((t_{i+1}-t_i)/2)}{(t_{i+1}-t_i)/2} \geq \frac{\sin(\theta/2)}{\theta/2}$. $\qquad\square$

Lemma 2. *For any r and s, $r < s < r + \pi$, in the domain of γ, the angle between $\gamma'(r)$ and the ray $\gamma(r)\gamma(s)$ is at most $\frac{s-r}{2}$.*

Proof. Assume again that $r = 0$ and that $\gamma(0) = O$; also assume w.l.o.g. that $\gamma'(0)$ is horizontal ($\gamma'(0) = (1,0)$). Let $\gamma(s) = (x(s), y(s))$, and let $k(s) = y(s)/x(s)$ be the slope of the ray $O\gamma(s)$ (Fig. 2, left). Then the lemma is equivalent to $k(s) \leq \tan \frac{s}{2}$, which we will prove by showing that $k' \leq \frac{1 - \cos s}{\sin^2 s} = \frac{1}{2\cos^2(s/2)} = (\tan \frac{s}{2})'$.

By definition, for any $\tau < s$ the angle between $\gamma'(\tau)$ and $\gamma'(s)$ is at most $s - \tau$; in particular $\gamma'(s) \leq s$. It follows that $x'(s) = (x'(s), y'(s)) \cdot (1, 0) = \gamma'(s) \cdot \gamma'(0) = \cos \angle s0 \geq \cos s$, and thus $x(s) \geq \sin s$. Next, consider the unit vector $\eta = (y'(s), -x'(s))$, orthogonal to $\gamma'(s)$ (Fig. 2). Since the angle between $\gamma'(\tau)$ and η is at least $\pi/2 - (s - \tau)$, it follows that $(x'(\tau), y'(\tau)) \cdot \eta \leq \cos(\pi/2 - (s - \tau))$, or $x'(\tau)y'(s) - y'(\tau)x'(s) \leq \sin(s - \tau)$. Integrating over τ from 0 to s, we get $x(s)y'(s) - y(s)x'(s) \leq 1 - \cos s$. Combining this with $x(s) \geq \sin s$, we obtain what we need: $k' = (y/x)' = \frac{y'x - x'y}{x^2} \leq \frac{1 - \cos s}{\sin^2 s}$. $\qquad\square$

Corollary 2. *The angle between the ray $\gamma(t_{i-1})\gamma(t_i)$ and $\gamma'(t_i)$ is at most $\sin^{-1}\left(\frac{\min\{d_\theta, |\gamma(t_{i-1})\gamma(t_i)|\}}{2}\right)$*

Proof. By the lemma, the angle between the ray $\gamma(t_{i-1})\gamma(t_i)$ and $\gamma'(t_i)$ is at most $\frac{t_i - t_{i-1}}{2}$, which is always at most $\theta/2 = \sin^{-1}(\frac{d_\theta}{2})$. So, it suffices to consider the case where $|\gamma(t_{i-1})\gamma(t_i)| < d_\theta$. But in this case, $\frac{t_i - t_{i-1}}{2} \leq \sin^{-1}(\frac{|\gamma(t_{i-1})\gamma(t_i)|}{2})$, by Lemma 1. $\qquad\square$

In summary, we have shown the following:

Theorem 1. *If γ is any cc-path and $\hat{\gamma}$ any θ-discretization of γ, then (i) $\hat{\gamma}$ is a θ-bcc-path, and (ii) $|\gamma|\frac{\sin(\theta/2)}{\theta/2} \leq |\hat{\gamma}| \leq |\gamma|$.*

Proof. (i) That $\hat{\gamma}$ is a θ-bcc-path is an immediate consequence of Corollary 2, since the angle between the ray $\gamma(t_{i-1})\gamma(t_i)$ and the ray $\gamma(t_i)\gamma(t_{i+1})$ is just the sum of the angles formed by these rays with $\gamma'(t_i)$.

(ii) It is clear that the length of any θ-discretization of a smooth curve γ is no greater than the length of γ. On the other hand, it follows immediately from Corollary 1 that its length cannot be less than $|\gamma|\frac{\sin(\theta/2)}{\theta/2}$.   □

The fact that the bounds on $|\hat{\gamma}|$ coincide in the limit as θ approaches zero, will be used to obtain properties of shortest smooth paths as a limit of the properties of their discrete counterparts.

Remark 2. It is worth noting at this point that our definition of θ-dcc-path, because of its "local" nature, rules out some paths that may be seen as having bounded curvature. For example, a "sawtooth" approximation of a straight line (see Fig. 3) does not qualify as a θ-dcc-path if the pitch of the teeth (turn angle), no matter how small, is too sharp relative to the size of the teeth (edge length).

Configurations. A *configuration* is a pair (p, ϕ), where p is a point and ϕ is a direction (unit vector). We say that a polygonal path $P = \langle v_1, v_2, \ldots, v_n \rangle$ satisfies endpoint configurations (v_1, ϕ_1) and (v_n, ϕ_n) if (i) the difference between the angle of the ray $v_1 v_2$ and ϕ_1, is no more than $\sin^{-1}(\frac{\min\{d_\theta, |v_1 v_2|\}}{2})$, and (ii) the difference between the angle of the ray $v_{n-1}v_n$ and ϕ_n, is no more than $\sin^{-1}(\frac{\min\{d_\theta, |v_{n-1}v_n|\}}{2})$.

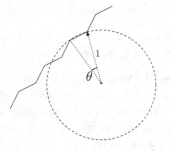

Remark 3. This is equivalent to asserting that the path $\langle v_0, v_1, v_2, \ldots, v_n, v_{n+1} \rangle$, formed from P by adding an arbitrarily short edge of direction ϕ_1 (respectively, ϕ_n) to the start (respectively, end) of P, has bounded discrete-curvature.

Fig. 3. A sawtooth path that does *not* qualify as a θ-dcc-path.

Remark 4. It is easy to confirm that, for *any* intermediate configuration (v_i, ϕ_i), the composition of any dcc-path $\langle v_1, v_2, \ldots, v_i \rangle$ that satisfies endpoint configurations (v_1, ϕ_1) and (v_i, ϕ_i) with any dcc-path $\langle v_i, v_{i+1}, \ldots, v_n \rangle$ that satisfies endpoint configurations (v_i, ϕ_i) and (v_n, ϕ_n) produces a dcc-path $\langle v_1, v_2, \ldots, v_n \rangle$ that satisfies endpoint configurations (v_1, ϕ_1) and (v_n, ϕ_n). Furthermore, if $P= \langle v_1, v_2, \ldots, v_n \rangle$ is any dcc-path that satisfies endpoint configurations (v_1, ϕ_1) and (v_n, ϕ_n), then, for all i, $1 < i < n$, there exists a direction ϕ_i such that (i) the sub-path $\langle v_1, v_2, \ldots, v_i \rangle$ is a dcc-path with endpoint configurations (v_1, ϕ_1) and (v_i, ϕ_i), and (ii) the sub-path $\langle v_i, v_{i+1}, \ldots, v_n \rangle$ is a dcc-path with endpoint configurations (v_i, ϕ_i) and (v_n, ϕ_n). On the other hand, breaking P at an arbitrary point in the interior of one of its edges may produce a path that no longer has bounded discrete-curvature (at the breakpoint).

We will be interested in characterizing shortest dcc-paths that satisfy specified endpoint configurations:

Definition 2. *A discrete-geodesic joining endpoint configurations (v_1, ϕ_1) and (v_n, ϕ_n), is a dcc-path that (i) satisfies the endpoint configurations (v_1, ϕ_1) and (v_n, ϕ_n), and (ii) has minimum total length among paths satisfying (i).*

Remark 5. It is by no means obvious that discrete-geodesics exist for all endpoint configurations. Dubins' proof [15] of the existence of smooth geodesics makes use of tools from functional analysis (in particular, Ascoli's theorem); in Sect. 4, we describe an alternative approach to establishing the existence of discrete-geodesics, having established a suitable characterization of the form the discrete-geodesics must take (if they exist).

Remark 6. It is straightforward to confirm that if a path contains a pair of successive edges $v_{i-1}v_i$ and v_iv_{i+1} whose *shortcut*, edge $v_{i-1}v_{i+1}$, has length at most d_θ, then the path can be made both shorter and smaller (fewer edges) by replacing the edges by their shortcut, without violating the curvature constraint. It follows that for the purposes of characterizing discrete-geodesics, we can restrict our attention to paths with finitely many edges formed by maximal[2] discrete circular arcs connected by (possibly degenerate) line segments. This assertion, which follows immediately from our definition, is the discrete analogue of a nontrivial property of smooth curvature-bounded geodesics, proved as Proposition 13 in [15].

1.1 Related Work

The books [26,27] provide general references that discuss curvature-constrained path planning in the broader context of non-holonomic motion planning. We note that study of curvature-constrained path planning has a rich history that long predates and goes well beyond robot motion planning, see for example the work of Markov [29] on the construction of railway segments.

The Dubins characterization of smooth geodesics has been rederived using techniques from optimal control theory in [7,40]. Variations and generalizations of the problem are studied in [6,8–10,12–14,19,28,30–34,37–39]. In addition, Dubins' characterization plays a fundamental role in establishing the existence as well as the optimality of curvature-constrained paths. Jacobs and Canny [23] showed that even in the presence of obstacles it suffices to restrict attention to paths of Dubins form between obstacle contacts and that if such a path exists then the shortest such path is well-defined. Fortune and Wilfong [20] give a super-exponential time algorithm for determining the existence of, but not actually constructing, such a path. Characterizing the intrinsic complexity of the existence problem for curvature-constrained paths is hampered by

---

[2] We ignore for the present the fact that successive maximal discrete circular arcs of opposite orientation could share an edge. In this case we are free to impose disjointness of arcs by assigning the shared edge to just one of the two arcs.

the fact that there are no known bounds on the minimum length or *intricacy* (number of elementary segments), expressed as a function of the description of the polygonal domain, of obstacle-avoiding paths in Dubins form. In a variety of restricted domains polynomial-time algorithms exist that construct shortest bounded-curvature paths [1,2,4].

A discretization of curvature-constrained motion was studied by Wilfong [42,43]. However, his setting is different from ours since he considered discretized *environment*, and not discrete paths. A practical way of producing paths with length and turn constraints is presented in [41]. For some other recent work on bounded-curvature paths see [3,5,11,16,17,21,22,24].

1.2 Our Approach

We study properties of discrete-geodesics that are free of inflections, as well as their smooth counterparts. In Sect. 2 we motivate the study of this restricted class of paths by proving that unrestricted discrete-geodesics never need more than two internal inflections; i.e. all discrete geodesics are formed by the concatenation of at most three inflection-free discrete-geodesics.

Section 3 establishes the central result of the paper: a precise characterization of the form of all discrete-geodesics (if they exist). In Sect. 4 we use this characterization to outline a proof of a characterization of smooth inflection-free geodesics (establishing, by a simple limiting argument, this interesting variant of the Dubins characterization). We also include a simple geometric proof of the existence of one important special case of discrete-geodesics that illustrates the strength of our characterization.

We note that similar methods, proving properties of smooth curves using discretization, were already used by Schur in his paper of 1921; interestingly, exactly these problems, considered by Schur [36] and Schmidt [35], led Dubins to his result.

2 Inflections in Discrete-Geodesics

We now start our investigation of the structure of discrete-geodesics. An edge e of a polygonal path is an *inflection* edge if the edges adjacent to e lie on the opposite sides of (the supporting line of) e. Such an edge is said to have *positive inflection* if the path makes a left turn into and a right turn out of e (and *negative inflection*, otherwise). Note that, in accordance with our interpretation of endpoint conditions as a zero-length edge of specified orientation, the first and last edges of a path are possible inflection edges. When we want to distinguish such edges, we refer to them as *endpoint inflection edges*; other inflection edges are referred to as *internal inflection edges*.

It is not hard to see that a dcc-path can have arbitrarily many inflection edges (of arbitrary lengths). However, minimum length such paths, can have no more than two internal inflection edges (of any length) *in total*.

2.1 More than Two Internal Inflection Edges is Impossible

Our first observation is that in any discrete-geodesic there can be at most one internal inflection edge of each turn type.

Lemma 3. *Any θ-dcc-path containing two or more internal inflection edges of the same type can be replaced by another θ-dcc-path, with fewer edges, whose total length is no longer than the original. In fact, if the inflection edges are non-parallel, the replacement path is strictly shorter.*

Proof. Let $P = \langle v_1, \ldots, a, b, c, d, \ldots, w, x, y, z, \ldots, v_n \rangle$ and suppose that both bc and xy are internal inflection edges with positive inflection (i.e. P turns left at both b and x and right at both c and y; see Fig. 4). Note that since edges bc and xy are internal inflection edges, the edges ab, cd, wx and yz all have strictly positive length.

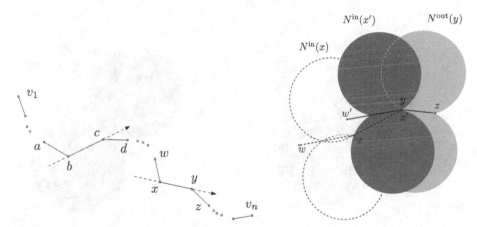

Fig. 4. A discrete curvature-constrained path with two positive inflection edges bc and xy.

Fig. 5. Full translation

We assume, without loss of generality, that the ray from b through c and the ray from x through y are either parallel or diverge (as illustrated). Then, any transformation of P that results from a translation of the sub-path between c and x along the vector xy (taking $c, \ldots x$ to $c', \ldots x'$) reduces the length of P (except when the inflection edges are parallel, in which case the length of P is preserved) and maintains the dcc-path property at b and c (since the edge bc lengthens, while the turns at both b and c are not increased).

Consider the situation when x has been translated all the way to y (see Fig. 5). The dcc-path property holds for the resulting path as long as $\operatorname{pred}(x')$ lies in $N^{out}(y)$ (or, equivalently, $\operatorname{succ}(y)$ lies in $N^{in}(x')$), where x' denotes the translation of x, etc. In this case, there is nothing left to prove since the path has one fewer edge (namely xy) than P. Hence we can assume that, after this full translation, $\operatorname{pred}(x')$ lies in $N^{out}(y)$.

If $\mathrm{pred}(x')$ lies in the right component of $N^{out}(y)$ (see Fig. 6) then it must lie in the segment of this circle cut off by the line through x and y, which implies that $|\mathrm{pred}(x')x'| < d_\theta$ and so $\mathrm{pred}(x') = w'$. It follows that when the translation is taken just to the point where w' lies on the boundary of $N^{out}(y)$, at which point y must still lie outside $N^{out}(\mathrm{pred}(x'))$ (see Fig. 7), we have $|w'y| < d_\theta$ and thus if we replace the edges $w'x'$ and $x'y$ at this point by the edge $w'y$, we must have a path that satisfies the dcc-path property.

Similarly, if $\mathrm{pred}(x')$ lies in the left half of $N^{out}(y)$ (see Fig. 6) then $\mathrm{succ}(y)$ must lie in the left half of $N^{in}(x')$ and in fact in the segment of this circle cut off by the line through x and y. As before, this implies that $|\mathrm{succ}(y)y| < d_\theta$ and so $\mathrm{succ}(y) = z$. It follows that when the translation is taken just to the point where z lies on the boundary of $N^{in}(x')$, at which point x' has not yet entered the interior of $N^{out}(y)$ (see Fig. 7), we have $|x'z| < d_\theta$ and thus if we replace the edges $x'y$ and yz at this point by the edge $x'z$, we must have a path that satisfies the dcc-path property.

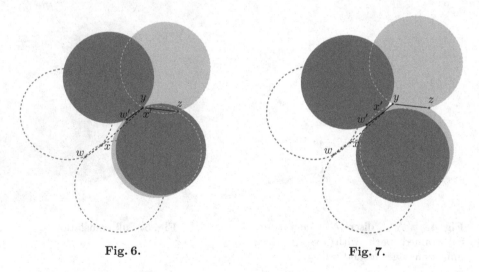

Fig. 6.   Fig. 7.

Since in both of these remaining cases the resulting path has one fewer edge than P, the result follows. Note that the only situation where the path has not had its length strictly reduced is where the inflection edges are parallel (and so the translation is length preserving). $\square$

Remark 7. It is worth observing at this point that Lemma 3 applies as well to the case in which the inflection edge xy is an endpoint inflection. In this case, we are not able to conclude that the transformed path has one fewer edge (since edge yz has length zero), but it does follow from our argument that if P cannot be shortened then the endpoint inflection edge must be a chord of the circle that defines the endpoint configuration at y.

It follows from Lemma 3 that, ignoring the length zero edges at the path endpoints, any discrete-geodesic is the concatenation of at most three inflection-free

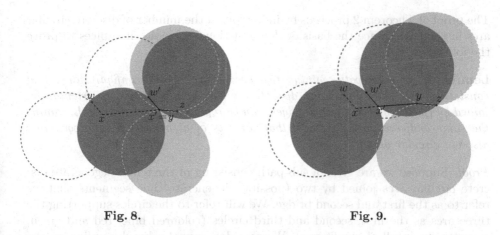

Fig. 8.   Fig. 9.

sub-paths. This motivates our next focus on the form of inflection-free discrete-geodesics.

3   A Normal Form for Inflection-Free Geodesics

We have already noted that any short-est dcc-path of minimum size consists of a finite number of maximal discrete circular arcs connected by (possibly degenerate) line segments that we refer to as *bridges*. Here we include the (possibly degenerate) circular arcs (e.g. vertex k in Fig. 10) supported by the circles (shown as dashed) that define the endpoint configurations of the path. A bridge vw is *degenerate* if $v = w$ (e.g. the first bridge, vertex d, in Fig. 10). Of course, if a given path has no inflection edges, all of the discrete circular arcs have the same ori-

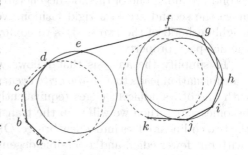

Fig. 10. A discrete-curvature-constrained path with five discrete arcs (including one degenerate arc) and no inflections.

entation; without loss of generality we will assume that they are all clockwise oriented (see Fig. 10).

The main result of this section is the following theorem. It amounts to a special case of a general characterization theorem for discrete-geodesics and is a fundamental building block for the proof of that theorem.

Theorem 2. *Any inflection-free discrete-geodesic, joining two specified endpoint configurations, is composed of a sequence of at most four discrete circular arcs, at most two of which are non-degenerate, joined by (possibly degenerate) bridges.*

The proof of Theorem 2 proceeds by induction on the number of discrete circular arcs in the path. Since the basis of the induction is obvious, it suffices to prove the following:

Lemma 4. *Any dcc-path, joining two specified endpoint configurations, that consists of three non-degenerate discrete circular arcs of the same orientation, joined by (possibly degenerate) bridges, can be replaced by a shorter path, joining the same endpoint configurations, that consists of at most two non-degenerate discrete circular arcs.*

Proof. Suppose we are given a sub-path consisting of three non-degenerate discrete circular arcs joined by two (possibly degenerate) line segments that we refer to as the first and second bridge. We will refer to the circles supporting the three arcs as the first, second and third circles (coloured blue, red and green, respectively, in all of our figures). We consider several cases depending on the nature of the two bridges (degenerate or not) and the total turn of the second arc (essentially whether or not it exceeds π). With only one explicitly noted exception, we use one of two continuous *shortening transformations* both of which involve moving all (or most) of the vertices on the second arc while keeping the other arcs fixed: (i) a *pivot* rotates all (except possibly the opposite endpoint) of the vertices on the second arc about one of the arc endpoints; and (ii) a *slide* translates all (except possibly the opposite endpoint) of the vertices on the second arc along one of the non-degenerate bridges. Since both transformations move the second arc in a rigid fashion, we need only consider vertices in the neighbourhood of the two bridges to confirm that the transformations preserve the dcc-path property.

To simplify the analysis that follows we will assume throughout that if a transformation leads to a *co-linearity event*: one of the bridges becomes co-linear with one of its adjacent edges (equivalently, the turn at some bridge endpoint becomes zero), then we will stop the transformation at this point and combine the two co-linear edges into a new bridge. Obviously, this results in a simpler path (with one fewer edge) and a possible degeneration of one of the discrete circular arcs. With this exception, all of our transformations terminate with either (i) the second arc becoming co-circular with the first or third (a *co-circularity event*), in which case a bridge has been eliminated, (ii) a formerly non-degenerate bridge becoming degenerate (a *bridge degeneration event*), or (iii) a bridge intersecting one of its associated circles in a chord of length d_θ (a *maximal chord event*). In the second event, the resulting path is simpler in the sense that a path with a narrow second arc (total turn less than π) gets measurably narrower, and one with a wide second arc (total turn greater than π) gets measurably wider. The third event is treated differently, depending on the intersection of the bridge with it second associated circle. In all cases, the transformations are easily seen to not only *shorten* the path, they also arguably leave it in a form that is *simpler* than it was to start, from which it immediately follows that the full reduction consists of only finitely many transformation steps.

Case I: both bridges are non-degenerate. We begin by considering the case where both bridges are non-degenerate. As we shall see, if one or more of the bridges is degenerate, a shortening transformation exists that will bring us back to this case.

There are two sub-cases to consider. In the first sub-case the turn from the first bridge edge to the second is less than or equal to π (see Fig. 11). First note that if we slide the middle discrete arc (vertices c through x) along the first bridge edge (taking c towards b) we maintain the discrete bounded curvature property at b and c as long as b (respectively, c) lies outside the second circle (respectively, first) circle (i.e. until the bridge bc becomes degenerate). Meanwhile, the discrete bounded curvature property is maintained at the endpoints of the second bridge edge (xy) as long as the predecessor (r) of the outer point (y) lies outside of the third circle (because of the direction of the translation the successor of the other bridge endpoint point (x) cannot enter the second circle). If this point r meets the third circle (a maximal chord event) while outside of the second circle, then point r can replace y as the outer point of the second bridge (leading to a shortening of the second discrete arc), and we can continue in Case I.

By symmetry the analogous properties hold if we slide the middle discrete arc along the second bridge edge (taking x to y). Since both of these translations serve to shorten the curve, we can assume that they have been done until either or both of the bridges have degenerated (taking us to Case II or Case III below) or we are left with unresolved maximal chord events on both bridges. In the latter case, the successor point of b (illustrated by p in Fig. 12) must lie on the first circle, the predecessor point of y (illustrated as r) must lie on the third circle, and both p and r must lie inside the second circle.

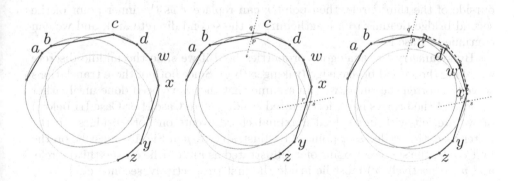

Fig. 11.   Fig. 12.   Fig. 13.

To deal with this last situation, we observe that either (i) the distance $|rx|$ must be at least the distance from p to the point q on the second circle intersected by the line through p with the slope of the second bridge edge, or (ii) the distance $|pc|$ must be at least the distance from r to the point s on the second circle intersected by the line through r with the slope of the first bridge edge. (It is easily confirmed that if neither of these hold, we get a contradiction of the fact that the slope of edge qc must exceed the slope of edge xs.) Without loss of generality, we assume that the first of these holds. Then, if we break edge bc at point p and translate the second arc along the second bridge edge, the second circle (specifically point q) must meet point p before x reaches r (note that, since $|ry| = d_\theta$, x reaches r before y enters the second circle). If we stop the translation at this point (see Fig. 13) we see that the first bridge has been replaced by two edges (bp and pc) which become part of the first and second discrete arc respectively; i.e. vertex p is a degenerate bridge, taking us to Case II. (It is worth noting here that as the translation takes q to p the discrete bounded curvature property is initially violated at c. It is restored just when q coincides with p. This is the reason why we need to ensure that the bridge remains feasible for vertices x and y until q reaches p. It is not at all clear that a transformation exists that is guaranteed to shorten the path in the situation under consideration, while preserving the discrete bounded curvature property throughout.)

The second sub-case, where the turn from the first bridge edge to the second is greater than π (see Fig. 14), is treated in a very similar fashion. As in the first sub-case, we note that if we slide the middle discrete arc along the first bridge edge (taking c towards b) we maintain the discrete bounded curvature property at b and c as long as b (respectively, c) lies outside the second (respectively, first) circle (i.e. until the first bridge becomes degenerate). Meanwhile the discrete bounded curvature property is maintained at the endpoints of the second bridge edge (xy) as long as the successor (r) of the inner point (x) lies outside of the second circle (because of the direction of the translation the predecessor of the other bridge endpoint (y) cannot enter the third circle). If this point r meets the second circle (a maximal chord event) while outside of the third circle, then point r can replace x as the inner point of the second bridge (leading to a lengthening of the second discrete arc), and we can continue in Case I.

By symmetry the analogous properties hold if we slide the middle discrete arc along the second bridge edge (taking x to y). Since both of these translations serve to shorten the curve, we can assume that they have been done until either or both of the bridges have degenerated (taking us to Case II or Case III below) or we are left with unresolved maximal chord events on both bridges. In the latter case, the predecessor point of c (illustrated by p in Fig. 15) must lie on the first circle, the successor point of x (illustrated as r) must lie on the third circle, and p (respectively, r) must lie inside the first (respectively, second) circle.

To deal with this last situation, we observe that either (i) the distance $|ry|$ must be at least the distance from p to the point q on the circle associated with the first arc intersected by the line through p with the slope of the second bridge edge, or (ii) the distance $|pb|$ must be at least the distance from r to the point s on the circle associated with the third arc intersected by the line through r with the slope of the first bridge edge. (As before, it is easily confirmed that if neither of these hold, we get a contradiction of the fact that the slope of edge bq must exceed the slope of edge ys.) Assume, without loss of generality that

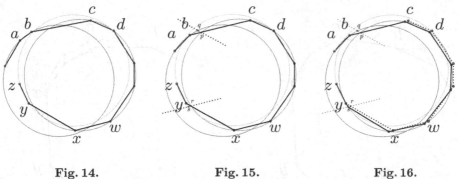

Fig. 14.   Fig. 15.   Fig. 16.

the second of these holds. Then, if we break edge xy at point r and translate the second arc, including the point r, along the first bridge edge, the point r must encounter the circle associated with the third arc before p reaches b. If we stop the translation at this point (see Fig. 16) we see that the second bridge has been replaced by two edges (xr and ry) which become part of the second and third discrete arc respectively; i.e. vertex r is a degenerate bridge, taking us to Case II. (As before, we note that the translation produces a path that violates the discrete bounded curvature property initially, but it is restored just when s coincides with r.)

Case II: one bridge is degenerate and the other is not. We assume, without loss of generality, that the first bridge is degenerate. There are two sub-cases again that depend on the span of the second arc. In the first sub-case (see Fig. 17) the total turn from the first edge (bc) after the degenerate bridge (b) to the second bridge edge (xy) is less than or equal to π. If we translate the middle discrete arc, excluding the first bridge point, (i.e. the vertices c through x) along the second bridge edge (taking x towards y) (see Fig. 18), we maintain the discrete bounded curvature property until the first of two events occurs: (i) x (respectively, y) joins the third (respectively, second) circle, or (ii) the successor (c) of the degenerate bridge (b) joins the first circle. The first event coincides with the degeneration of the second bridge, while the second event leaves us with a new degenerate first bridge (vertex c), and hence a new instance of Case II with a smaller second arc.

In the second sub-case (see Figs. 19 and 21), where the total turn from the first edge (bc) after the degenerate bridge (b) to the second bridge edge (xy) is greater than π, we again slide the middle discrete arc, this time including the first bridge point, (i.e. the vertices b through x) along the second bridge edge (taking x towards y). The discrete bounded curvature property is maintained at x and y unless x (respectively, y) joins the third (respectively, second) circle (i.e. the second bridge becomes degenerate). Meanwhile, if b moves outside of the first circle (see Fig. 20), then the discrete bounded curvature property is maintained at a and b until a joins the second circle, at which point a replaces b as a degenerate bridge, so we can continue in Case II. Alternatively, if b moves inside the first circle, we maintain feasibility of the transformed path by (continuously)

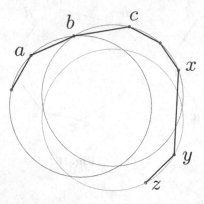

Fig. 17.

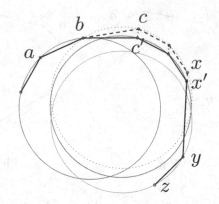

Fig. 18.

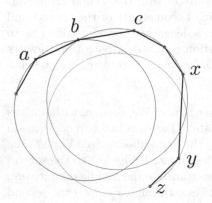

Fig. 19.

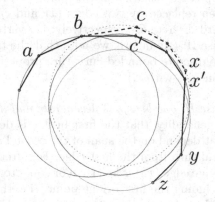

Fig. 20.

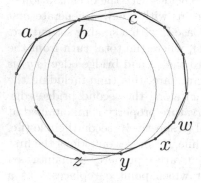

Fig. 21.

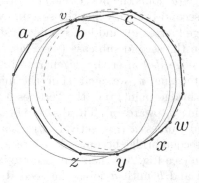

Fig. 22.

replacing the bridge point by the intersection point v of the first two circles (see Fig. 22). It is straightforward to confirm that both $|av| < |ab|$ and $|vc'| < |bc|$. Thus, the transformation can continue until v coincides with either a or c, at which point either a or c becomes a degenerate bridge, and we can continue in Case II, with one fewer edge.

Case III: both bridges are degenerate. As in both previous cases, there are two sub-cases depending on the span of the middle arc. If the middle arc spans less than a half circle (refer to Fig. 23) then we can transform the path by rotating the second arc, excluding the first bridge point, counterclockwise about the second bridge point (see Fig. 24). Of course, if this rotation continues long enough the second and third circle will coincide, at which point the second bridge disappears. Prior to this, the transformation preserves the length of all edges except for the first edge after the first bridge point (bc in Fig. 24) which shortens, since the distance from both endpoints of this edge to the second bridge point (y) is unchanged but the angle they form with the second bridge point decreases. The transformation continues until the first vertex on the second arc (c) meets the first circle, at which point it becomes a new degenerate first bridge. We then continue in Case III, with a smaller second arc.

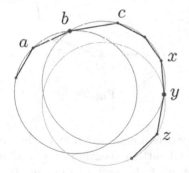 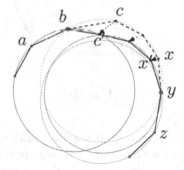

Fig. 23.   Fig. 24.

In the second sub-case, the middle arc spans at least a half circle (refer to Figs. 25 and 27). Then we can transform the path by rotating the second arc, this time including both bridge points, counterclockwise about the second bridge point. As in the previous sub-case, if this rotation continues long enough the second and third circle will coincide at which point the second bridge disappears. Prior to this, there are two cases to consider depending on the trajectory of the first bridge point (b).

If the first bridge point (b) moves outside the first circle (see Fig. 26), the discrete bounded curvature property is maintained at a and b until a joins the second circle, at which point either a or c replaces b as a degenerate bridge, so we can continue in Case III.

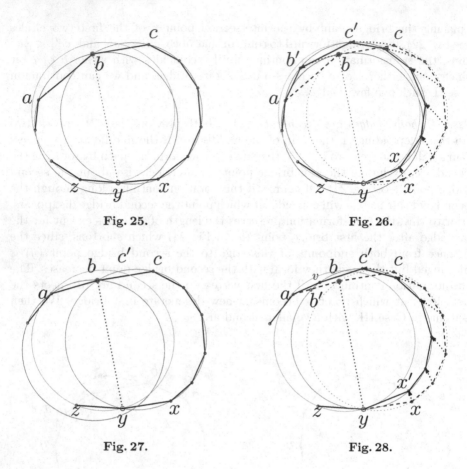

Fig. 25.   Fig. 26.

Fig. 27.   Fig. 28.

Alternatively, if the first bridge point (b) moves inside the first circle (see Fig. 28), we maintain feasibility of the transformed path by (continuously) replacing the bridge point by the intersection point v of the first two circles. It is straightforward to confirm that both $|av| < |ab|$ and $|vc'| < |bc|$. Thus, the rotation can continue until v coincides with either a or c, at which point either a or c becomes a degenerate bridge, and we can continue in Case III, with one fewer edge.   □

4   Existence and Uniqueness of ς-geodesics

Careful inspection of the proof of Lemma 4 shows that it applies even when the first or third discrete circular arc is degenerate (i.e. arises from an endpoint constraint), provided the second discrete circular arc spans at most a half circle. Furthermore, if the second circular arc in some locally shortest path spans more than a half circle, then if the first (or third) arc is degenerate, it must be the case that the path starts (or ends) with an edge that (i) is an extension of

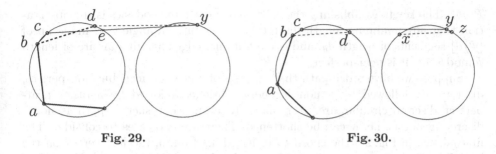

Fig. 29.   Fig. 30.

the corresponding endpoint configuration and (ii) cuts the middle circle with a maximal chord.

Taking this into consideration, Theorem 2 can be strengthened to provide a very tight characterization of the *form* of inflection-free θ-discrete-geodesics, if they exist: they are formed by two (or fewer) θ-discrete circular arcs of the same orientation, joined by a (possibly degenerate) bridge, and preceded and followed by (possibly degenerate) edges that are extensions of the endpoint configurations. Furthermore, when the extension of one endpoint configuration is non-degenerate, either (i) there is only a single non-degenerate θ-discrete circular arc, or (ii) the extension of the other endpoint configuration must be degenerate, and the adjacent θ-discrete circular arc must span more than a half circle.

We note that, as θ goes to zero, this refined characterization of inflection-free θ-discrete-geodesics describes a family of smooth geodesics, including the sole inflection-free geodesic specified by Dubins' general characterization. In this way, we can derive an analogue of Dubins' result for inflection-free geodesics. Uniqueness, in the smooth case, is a direct consequence of the uniqueness of their discrete counterparts, together with our discretization theorem (Theorem 1).

Clearly paths of the form specified by Theorem 2, joining specified endpoint configurations, always exist. To argue the existence of discrete-geodesics, it remains to argue that the infimum of the lengths of paths of this form is always realized by a path of this form. It would suffice to use a compactness argument (of the style used by Dubins), but it turns out to be both simpler and more revealing to argue this geometrically. We will do so for general (not necessarily inflection-free) discrete-geodesics in a companion paper. To give some sense of the kind of arguments involved, we consider just one special case here: an inflection-free dcc-path is *endpoint-anchored* if it is formed by two θ-discrete circular arcs respecting the two endpoint constraints, joined by a bridge of length at least $2d_\theta$. (Note that such paths correspond to the unique inflection-free paths in Dubin's characterization, in the limit as θ goes to zero.)

To this end, we say that a θ-discrete arc consisting of a sequence of edges of length exactly d_θ is *perfect*, and a θ-discrete arc consisting of a sequence of edges all but one of which have length exactly d_θ is *near-perfect*. With this, we can assert that:

Claim 3. *Endpoint-anchored geodesics exist and are composed of two perfect discrete circular arcs joined by a non-degenerate bridge.*

Proof. The key to establishing the existence of discrete-geodesics is the observation that any minimum length discrete arc spanning an angle Ψ is made up of $\lfloor \Psi/\theta \rfloor$ segments of length d_θ, and one additional edge spanning an arc of length $\Psi \bmod \theta$, i.e. it is near-perfect.

Suppose we have a dcc-path that consists of a near-perfect, but non-perfect, discrete arc followed by a non-degenerate bridge, followed by another near-perfect discrete circular arc. It remains to argue that such a path is not a discrete-geodesic, i.e. it can be shortened. There are two cases to consider. The first, shown in Fig. 29, the chord bd of length d_θ from b, the last vertex on the initial perfect arc, crosses the bridge cy. In this case, it is straightforward to show (since $|bd| + |de| \leq |bc| + |ce|$ and $|dy| < |de| + |ey|$) that replacing sub-path bcy by bdy must produce a shorter dcc-path with the same endpoints.

Alternatively, we can assume (see Fig. 30) that neither bd nor the corresponding chord xz on the second circle cross the bridge cy. In this case, if we pivot the bridge cy about its endpoint y then the dcc-path property is preserved until the bridge hits the first of b, d or x. But since this transformation leads to a shorter path in all three situations. $\qquad\square$

5 Conclusion

We introduced a discrete model of curvature-constrained motion and studied some of its properties, in particular the structure of geodesics in this model. Our focus here has been primarily on inflection-free paths, which we have demonstrated constitute an essential component of unrestricted geodesics. We have also illustrated the utility of our characterization in relating properties of smooth geodesics as the limiting case of our discrete geodesics.

In a subsequent paper we will extend our characterization of inflection-free discrete-geodesics to the general case, including a re-derivation of the full Dubins characterization of smooth geodesics, using similar limiting arguments. We believe that discrete versions of curvature-constrained motions that include reversals (cf. [31]) can be formulated in the same way.

Acknowledgements. We thank Sergey Bereg, Stefan Foldes, Irina Kostitsyna and Joe Mitchell for discussions.

References

1. Agarwal, P.K., Biedl, T., Lazard, S., Robbins, S., Suri, S., Whitesides, S.: Curvature-constrained shortest paths in a convex polygon. In: SoCG (1998)
2. Agarwal, P.K., Raghavan, P., Tamaki, H.: Motion planning for a steering-constrained robot through moderate obstacles. In: SToC, pp. 343–352 (1995)
3. Ahn, H.-K., Cheong, O., Matousek, J., Vigneron, A.: Reachability by paths of bounded curvature in a convex polygon. Comput. Geom. **45**(1–2), 21–32 (2012)
4. Bereg, S., Kirkpatrick, D.: Curvature-bounded traversals of narrow corridors.In: SoCG, pp. 278–287 (2005)

5. Bitner, S., Cheung, Y.K., Cook IV, A.F., Daescu, O., Kurdia, A., Wenk, C.: Visiting a sequence of points with a bevel-tip needle. In: López-Ortiz, A. (ed.) LATIN 2010. LNCS, vol. 6034, pp. 492–502. Springer, Heidelberg (2010)
6. Boissonnat, J.-D., Bui, X.-N.: Accessibility region for a car that only moves forwards along optimal paths. Research report 2181, INRIA Sophia-Antipolis (1994)
7. Boissonnat, J.-D., Cérézo, A., Leblond, J.: Shortest paths of bounded curvature in the plane. Int. J. Intell. Syst. **10**, 1–16 (1994)
8. Bui, X.-N., Souères, P., Boissonnat, J.-D., Laumond, J.-P.: Shortest path synthesis for Dubins nonholonomic robot. In: IEEE International Conference Robotics Automation, pp. 2–7 (1994)
9. Chang, A., Brazil, M., Rubinstein, J., Thomas, D.: Curvature-constrained directional-cost paths in the plane. J. Global Optim. **53**(4), 663–681 (2011)
10. Chitsaz, H., LaValle, S.: Time-optimal paths for a Dubins airplane. In: 46th IEEE Conference on Decision and Control (2007)
11. Chitsaz, H., Lavalle, S.M., Balkcom, D.J., Mason, M.T.: Minimum wheel-rotation paths for differential-drive mobile robots. Int. J. Rob. Res. **28**, 66–80 (2009)
12. Chitsaz, H.R.: Geodesic problems for mobile robots. Ph.D. thesis, University of Illinois at Urbana-Champaign, Champaign, IL, USA, AAI3314745 (2008)
13. Djath, K., Siadet, A., Dufaut, M., Wolf, D.: Navigation of a mobile robot by locally optimal trajectories. Robotica **17**, 553–562 (1999)
14. Dolinskaya, I.: Optimal path finding in direction, location and time dependent environments. Ph.D. thesis, The University of Michigan (2009)
15. Dubins, L.E.: On curves of minimal length with a constraint on average curvature and with prescribed initial and terminal positions and tangents. Am. J. Math. **79**, 497–516 (1957)
16. Duindam, V., Jijie, X., Alterovitz, R., Sastry, S., Goldberg, K.: Three-dimensional motion planning algorithms for steerable needles using inverse kinematics. Int. J. Rob. Res. **29**, 789–800 (2010)
17. Edison, E., Shima, T.: Integrated task assignment and path optimization for cooperating uninhabited aerial vehicles using genetic algorithms. Comput. Oper. Res. **38**, 340–356 (2011)
18. Eriksson-Bique, S.D., Kirkpatrick, D.G., Polishchuk, V.: Discrete dubins paths. CoRR, abs/1211.2365 (2012)
19. Foldes, S.: Decomposition of planar motions into reflections and rotations with distance constraints. In: CCCG'04, pp. 33–35 (2004)
20. Fortune, S., Wilfong, G.: Planning constrained motion. Ann. Math. AI **3**, 21–82 (1991)
21. Furtuna, A.A., Balkcom, D.J.: Generalizing Dubins curves: minimum-time sequences of body-fixed rotations and translations in the plane. Int. J. Rob. Res. **29**, 703–726 (2010)
22. Giordano, P.R., Vendittelli, M.: Shortest paths to obstacles for a polygonal dubins car. IEEE Trans. Rob. **25**(5), 1184–1191 (2009)
23. Jacobs, P., Canny, J.: Planning smooth paths for mobile robots. In: Li, Z., Canny, J.F. (eds.) Nonholonomic Motion Planning, pp. 271–342. Kluwer Academic Pubishers, Norwell (1992)
24. Kim, H.-S., Cheong, O.: The cost of bounded curvature. CoRR, abs/1106.6214 (2011)
25. Krozel, J., Lee, C., Mitchell, J.S.: Turn-constrained route planning for avoiding hazardous weather. Air Traffic Control Q. **14**, 159–182 (2006)
26. Latombe, J.-C.: Robot Motion Planning. Kluwer Academic Publishers, Boston (1991)

27. Li, Z., Canny, J.F. (eds.): Nonholonomic Motion Planning. Kluwer Academic Pubishers, Norwell (1992)
28. Ma, X., Castan, D.A.: Receding horizon planning for Dubins traveling salesman problems. In: Proceedings of the 45th IEEE Conference on Decision and Control, San Diego, CA, USA (2006)
29. Markov, A.A.: Some examples of the solution of a special kind of problem on greatest and least quantities. Soobshch. Kharkovsk. Mat. Obshch. **1**, 250–276 (1887). In Russian
30. Morbidi, F., Bullo, F., Prattichizzo, D.: On visibility maintenance via controlled invariance for leader-follower dubins-like vehicles. In: IEEE Conference on Decision and Control, (CDC), pp. 1821–1826 (2008)
31. Reeds, J.A., Shepp, L.A.: Optimal paths for a car that goes both forwards and backwards. Pac. J. Math. **145**(2), 367–393 (1990)
32. Reif, J., Wang, H.: Non-uniform discretization for kinodynamic motion planning and its applications. In: Laumond, J.-P., Overmars, M. (eds.) Algorithms for Robotic Motion and Manipulation, pp. 97–112. A.K. Peters, Wellesley, MA (1997); Proceedings of 1996 Workshop on the Algorithmic Foundations of Robotics, Toulouse, France (1996)
33. Robuffo Giordano, P., Vendittelli, M.: The minimum-time crashing problem for the Dubins car. In: International IFAC Symposium on Robot Control SYROCO (2006)
34. Savla, K., Frazzoli, E., Bullo, F.: On the dubins traveling salesperson problems: novel approximation algorithms. In: Sukhatme, G.S., Schaal, S., Burgard, W., Fox, D. (eds.) Robotics: Science and Systems II. MIT Press, Cambridge (2006)
35. Schmidt, E.: Über das extremum der bogenlänge einer raumkurve bei vorgeschreibenen einschränkungen ihrer krümmung. Sitzber. Preuss. Akad. Berlin, pp. 485–490 (1925)
36. Schur, A.: Über die schwarzsche extremaleigenschaft des kreises unter den kurven konstanter krümmung. Math. Ann. **83**, 143–148 (1921)
37. Shkel, A.M., Lumelsky, V.J.: Classification of the dubins set. Robot. Auton. Syst. **34**(4), 179–202 (2001)
38. Sigalotti, M., Chitour, Y.: Dubins' problem on surfaces ii: nonpositive curvature. SIAM J. Control Optim. **45**(2), 457–482 (2006)
39. Sussman, H.J.: Shortest 3-dimensional paths with a prescribed curvature bound. In: Proceedings of 34th IEEE Conference Decision Control, pp. 3306–3311 (1995)
40. Sussmann, H.J., Tang, G.: Shortest paths for the Reeds-Shepp car: a worked out example of the use of geometric techniques in nonlinear optimal control. Research report SYCON-91-10, Rutgers University, New Brunswick, NJ (1991)
41. Szczerba, R.J., Galkowski, P., Glickstein, I.S., Ternullo, N.: Robust algorithm for real-time route planning. IEEE Trans. Aerosp. Electron. Syst. **36**, 869–878 (2000)
42. Wilfong, G.: Motion planning for an autonomous vehicle. In: Proceedings of IEEE International Conference Robotics Automation, pp. 529–533 (1988)
43. Wilfong, G.: Shortest paths for autonomous vehicles. In: Proceedings of 6th IEEE International Conference Robotics Automation, pp. 15–20 (1989)

Online Weight Balancing on the Unit Circle

Hiroshi Fujiwara[1]([✉]), Takahiro Seki[2], and Toshihiro Fujito[1]

[1] Toyohashi University of Technology, Toyohashi, Japan
{h-fujiwara,fujito}@cs.tut.ac.jp
[2] Computron Co. Ltd., Maebashi-shi, Japan
clithtel@gmail.com

Abstract. We consider a problem as follows: Given unit weights arriving in an online manner with the total cardinality unknown, upon each arrival we decide where to place it on the unit circle in $\mathbb{R}^2$. The objective is to set the center of mass of the placed weights as close to the origin as possible. We apply competitive analysis defining the competitive difference as a performance measure. We first present an optimal strategy for placing unit weights which achieves a competitive difference of $\frac{1}{5}$. We next consider a variant in which the destination of each weight must be chosen from a set of positions that equally divide the unit circle. We give a simple strategy whose competitive difference is 0.35. Moreover, in the offline setting, several conditions for the center of mass to lie at the origin are derived.

1 Introduction

Suppose that we are given a series of points, each with unit weight, one by one with the total cardinality unknown in advance. Our task is to place the points one by one on the unit circle in $\mathbb{R}^2$ while keeping a good balance. We are not allowed to move the point any more, once it is placed. The balance is measured by the Euclidean distance between the center of mass of the placed points and the origin.

The difficulty is that we do not know how many points will arrive in total. If we guess the total cardinality somehow at the beginning, then we may try to place the points, for example, in such a way that they equally divide the unit circle. If the guess is correct, the center of mass comes to the origin. However, if the guess fails, say, if one extra point arrives, we have to place it somewhere and then lose the good balance. Also in the case of fewer points than expected, we cannot achieve the balance as planned. In this paper we consider this problem from the viewpoint of *competitive analysis*.

Our Contribution. We apply competitive analysis adopting the *competitive difference* as a criterion of competitiveness of a strategy. The competitive difference is defined as the maximum difference between the cost incurred by the strategy and the cost incurred by an optimal offline strategy that knows the total cardinality of points in advance. Our results are summarized as follows:

© Springer International Publishing Switzerland 2014
J. Akiyama et al. (Eds.): JCDCGG 2013, LNCS 8845, pp. 65–76, 2014.
DOI: 10.1007/978-3-319-13287-7_6

(a) We present a non-trivial optimal strategy whose competitive difference is $\frac{1}{5}$. This means that according to our strategy, the cost is guaranteed to be at most the optimal offline cost plus $\frac{1}{5}$.

(b) We impose the n-cyclotomic constraint on the problem that for fixed n, the destination of each point has to be chosen from $\{(\cos \frac{2k\pi}{n}, \sin \frac{2k\pi}{n}) \mid 0 \leq k \leq n-1, k \in \mathbb{Z}\}$ and each position is occupied by at most one point. Depending on the parity of n, we give a simple and competitive strategy. Our strategy guarantees a competitive difference of 0.35 for odd n and $\frac{1}{3}$ for even n.

(c) We investigate the n-cyclotomic constrained problem in the offline setting, in which the cardinality of points is informed at the beginning. Even with the information of the cardinality, it is not clear whether there is a placement of points that lets the center of mass come exactly to the origin. We reveal several conditions for the existence of such a placement.

Related Work. To the best of our knowledge, this paper seems the first to focus on the placement of weighted objects that *arrive in an online manner* in terms of the optimal placement of their center of mass. One can find many studies with similar purposes in the offline setting: Kurebe et al. [6] considered the placement of weighted rectangles on $\mathbb{R}^2$ to let their center of mass approach the target position. Teramoto et al. [7] studied the insertion of points into the unit square in $\mathbb{R}^d$ in such a way that the Euclidean distance between any pair of points becomes as uniform as possible. Recently, Barba et al. [1] considered the problem that given a set of weights, a closed connected region, and a target position, we are asked to place the weights on the boundary of the region so that the center of mass lies at the target.

In consistent hashing, one can think that items and caching machines are both mapped to points on the unit circle [4,5]. In the context of the space science, satellite constellation design for covering the Earth's sphere has been of great interest, for example [3,9].

2   Problem Statement and Preliminaries

Throughout this paper a *point* denotes an individual object that is to be placed (or has been placed) on $\mathbb{R}^2$, while a *position* stands for where to place a point on $\mathbb{R}^2$. Each point has unit weight unless we specify otherwise. We sometimes identify a position on $\mathbb{R}^2$ and its xy coordinate, such as the origin $O = (0,0)$. $\overline{AB}$ denotes the Euclidean distance between the positions A and B.

We define the *online weight balancing problem* as follows. We are given a series of points, each with unit weight, in an online manner where the points arrive one by one and the total cardinality is unknown in advance. Our task is to place each point, upon its arrival, somewhere on the unit circle in $\mathbb{R}^2$. Once a point is placed, it cannot be moved any more. The objective is to minimize the *cost* which is defined as the Euclidean distance between the center of mass of the placed points and the origin, that is, the center of the unit circle.

A *strategy* for placing points is denoted by a sequence $\theta := (\theta_1, \theta_2, \ldots) \in S$ in the sense that it places the j-th point at $P_\theta(j) := (\cos \theta_j, \sin \theta_j)$, where S is the set of feasible strategies (specified later). The reason why a strategy is denoted thus simply is that any adaptive decision based on the history of the configuration does not help in this problem. When k points have arrived so far and been placed according to the strategy θ, the center of mass of the points lies at

$$G_\theta(k) := \left(\frac{1}{k} \sum_{j=1}^{k} \cos \theta_j, \frac{1}{k} \sum_{j=1}^{k} \sin \theta_j \right).$$

Then, the cost of the strategy θ is written as

$$C_\theta(k) := \overline{OG_\theta(k)} = \sqrt{\left(\frac{1}{k} \sum_{j=1}^{k} \cos \theta_j \right)^2 + \left(\frac{1}{k} \sum_{j=1}^{k} \sin \theta_j \right)^2}.$$

On the other hand, the *optimal offline cost*, that is, one with the cardinality known to be k in advance, is

$$C_{opt}(k) := \inf \{ C_\theta(k) \mid \theta \in S \}.$$

The performance of strategies for online problems is usually measured by the competitive ratio (see [2] for example), which would be defined as $\sup_{k \geq 1} \frac{C_\theta(k)}{C_{opt}(k)}$ for our problem. However, this is inconvenient here since $C_{opt}(k) = 0$ and $C_\theta(k) > 0$ happen often in the same time. We thus define and use the *competitive difference* instead. We say that the strategy θ has a competitive difference of d if

$$C_\theta(k) - C_{opt}(k) \leq d$$

holds for all $k \geq 1$. Apparently, $d \geq 0$. A smaller competitive difference means a better strategy.

In this paper we consider the online weight balancing problem under two different settings:

(A) The *basic problem*. We are allowed to place a point on an arbitrary position on the unit circle. Namely, the set of feasible strategies S is

$$\{ (\theta_1, \theta_2, \ldots) \mid 0 \leq \theta_j < 2\pi \text{ for } j \geq 1 \}.$$

(B) The *n-cyclotomic problem*. For fixed n, the destination of each point is chosen from $\{ (\cos \frac{2k\pi}{n}, \sin \frac{2k\pi}{n}) \mid 0 \leq k \leq n - 1, k \in \mathbb{Z} \}$, that is, a set of n positions that equally divide the unit circle into n arcs. Any position should not be occupied more than once. Formally, we set S to

$$\left\{ \left(\frac{2m_1\pi}{n}, \frac{2m_2\pi}{n}, \ldots, \frac{2m_n\pi}{n} \right) \; \middle| \; 0 \leq m_j \leq n - 1, m_j \in \mathbb{Z}^n \text{ for } 1 \leq j \leq n; \right.$$
$$\left. m_j \neq m_k \text{ for } 1 \leq j < k \leq n \right\}.$$

We assume in addition that the cardinality of the arriving points is at most n.

3 Basic Problem

3.1 Optimal Online Strategy

We first show a lower bound on the competitive difference and then give a strategy whose competitive difference coincides with that value. We begin by presenting a simple lemma on the offline cost.

Lemma 1. *For the basic problem, it holds that*

$$C_{opt}(k) = \begin{cases} 1, & k = 1; \\ 0, & k \geq 2. \end{cases}$$

Proof. It is trivial that $C_{opt}(1) = 1$ since the cost is one wherever we place a single point. For $k \geq 2$, just adopt the strategy $(0, \frac{2\pi}{k}, \frac{4\pi}{k}, \ldots, \frac{2(k-1)\pi}{k})$.   □

By rotational symmetry, we can assume that an optimal strategy satisfies $\theta_1 = 0$ and $0 \leq \theta_2 \leq \pi$. Let $\alpha := 2\arccos\frac{1}{5}$ ($\approx 157°$), which is a key angle for obtaining an optimal strategy. The next lemma gives a lower bound on the competitive difference.

Lemma 2. *Any strategy for the basic problem has a competitive difference of at least $\frac{1}{5}$.*

Proof. Fix a strategy $\boldsymbol{\theta}$ arbitrarily. By rotational symmetry, we can assume that $\theta_1 = 0$ and $0 \leq \theta_2 \leq \pi$. We will show that the competitive difference is at least $\frac{1}{5}$ regardless of the value of θ_2.
   (i) Case $0 \leq \theta_2 < \alpha$. We have

$$G_{\boldsymbol{\theta}}(2) := \left(\frac{1}{2}(1 + \cos\theta_2), \frac{1}{2}\sin\theta_2\right).$$

Since $x \mapsto \cos\frac{x}{2}$ is a decreasing function on $[0, \pi]$,

$$C_{\boldsymbol{\theta}}(2) = \frac{1}{2}\sqrt{(1 + \cos\theta_2)^2 + \sin^2\theta_2} = \cos\frac{\theta_2}{2} > \cos\frac{\alpha}{2} = \frac{1}{5}.$$

On the other hand, $C_{opt}(2) = 0$ by Lemma 1. Therefore, the competitive difference is greater than $\frac{1}{5}$.
   (ii) Case $\alpha \leq \theta_2 < \pi$. We evaluate the cost after the third point has been placed. For ease of analysis, we square the cost:

$$C_{\boldsymbol{\theta}}(3)^2 = \left(\frac{1}{3}\sum_{j=1}^{3}\cos\theta_j\right)^2 + \left(\frac{1}{3}\sum_{j=1}^{3}\sin\theta_j\right)^2$$

$$= \frac{2}{9}\sin\theta_2\sin\theta_3 + \frac{2}{9}\cos\theta_3\cos\theta_2 + \frac{2}{9}\cos\theta_2 + \frac{2}{9}\cos\theta_3 + \frac{1}{3}.$$

Let us think of $C_{\boldsymbol{\theta}}(3)^2$ as a function of θ_3 with a fixed parameter θ_2. By differentiating $C_{\boldsymbol{\theta}}(3)^2$ with respect to θ_3, we obtain

$$\frac{\partial C_{\boldsymbol{\theta}}(3)^2}{\partial \theta_3} = -\frac{2}{9}\sin\theta_3 + \frac{2}{9}\sin\theta_2\cos\theta_3 - \frac{2}{9}\sin\theta_3\cos\theta_2$$

$$= \frac{4}{9}\sin\left(\frac{\theta_2}{2} - \theta_3\right)\cos\frac{\theta_2}{2}.$$

This implies that when $\theta_3 = \frac{\theta_2}{2} + \pi$, the function $C_{\boldsymbol{\theta}}(3)^2$ achieves a minimum of

$$\frac{2}{9}\sin\theta_2\sin\left(\frac{\theta_2}{2} + \pi\right) + \frac{2}{9}\cos\left(\frac{\theta_2}{2} + \pi\right)\cos\theta_2 + \frac{2}{9}\cos\theta_2 + \frac{2}{9}\cos\left(\frac{\theta_2}{2} + \pi\right) + \frac{1}{3}$$

$$= \frac{1}{9}\left(1 - 2\cos\frac{\theta_2}{2}\right)^2.$$

(Geometrically speaking, the optimal position of the third point is the midpoint of the longer arc connecting $P_{\boldsymbol{\theta}}(1)$ and $P_{\boldsymbol{\theta}}(2)$.) Hence, for general θ_3, it holds that

$$C_{\boldsymbol{\theta}}(3) \geq \frac{1}{3}\left(1 - 2\cos\frac{\theta_2}{2}\right) = \frac{1}{3}\left(1 - \frac{2}{5}\right) = \frac{1}{5}.$$

Again by Lemma 1, we know $C_{opt}(3) = 0$. Therefore, the competitive difference is at least $\frac{1}{5}$. □

The strategy $\overline{\boldsymbol{\theta}}$ defined below turns out to be optimal. Note that the choice of placement for the fourth and later points is a matter of taste; any placement is acceptable as long as the resulting cost does not exceed $\frac{1}{5}$. Here we choose a placement for the fourth and later points so that the analysis is easy to handle. See Fig. 1.

$$\overline{\theta}_j = \begin{cases} 0, & j = 1; \\ \alpha, & j = 2; \\ \frac{\alpha}{2} + \pi, & j \text{ is odd}, j \geq 3; \\ \frac{\alpha}{2}, & j \text{ is even}, j \geq 4. \end{cases}$$

Lemma 3. $C_{\overline{\boldsymbol{\theta}}}(1) = 1$, and $C_{\overline{\boldsymbol{\theta}}}(k) \leq \frac{1}{5}$ for all $k \geq 2$.

Proof. $C_{\overline{\boldsymbol{\theta}}}(1) = 1$ is trivial. For ease of notation we write $P_{\overline{\boldsymbol{\theta}}}(\cdot)$ and $G_{\overline{\boldsymbol{\theta}}}(\cdot)$ simply as $P(\cdot)$ and $G(\cdot)$, respectively. Although the lemma can be proved by explicitly calculating the coordinate of $G(k)$ for general $k \geq 2$, we here give a simpler proof based on geometric arguments. Applying the strategy $\overline{\boldsymbol{\theta}}$, we calculate $G(2) = (\frac{1}{25}, \frac{2\sqrt{6}}{25})$ and $G(3) = (-\frac{1}{25}, -\frac{2\sqrt{6}}{25})$. (See Fig. 1.) It is thus observed that the origin O lies on the segment $G(2)G(3)$ and $\overline{OG(2)} = \overline{OG(3)} = \frac{1}{5}$. Therefore, the proof is done if $G(k)$ lies on the segment $G(2)G(3)$ for all $k \geq 2$.

We begin by proving that every $G(k)$ is on the line $G(2)G(3)$, not necessarily on that segment. Please note that $P(3) = P(5) = P(7) = \cdots = (-\frac{1}{5}, -\frac{2\sqrt{6}}{5})$ and $P(4) = P(6) = P(8) = \cdots = (\frac{1}{5}, \frac{2\sqrt{6}}{5})$ are on the line $G(2)G(3)$. For $k \geq 4$, $G(k)$

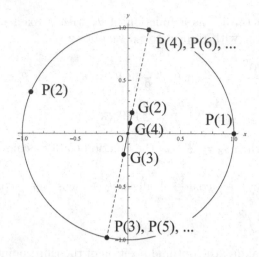

Fig. 1. Placement of points $P(1), P(2), \ldots$ according to the optimal online strategy $\overline{\theta}$ in Theorem 1 for the basic problem. $G(i)$ is the center of mass when i points have been placed so far.

can be calculated as the center of mass of a point with weight $k-1$ at $G(k-1)$ and one with weight unity at $P(k)$. Hence, $G(k)$ lies on the line $G(2)G(3)$ if $G(k-1)$ does so. Thus, we know inductively that every $G(k)$ is on the line $G(2)G(3)$.

We next show by induction that for odd $k \geq 2$, $G(k)$ is on the segment $OG(2)$. The claim is trivial for $k = 2$. Suppose that $G(k-2)$ lies on the segment $OG(2)$ for some odd $k(\geq 4)$. Consider that two points are added at $P(k-1)$ and $P(k)$ at once. The center of mass of these two points is obviously at the origin. Then, $G(k)$ is regarded as the center of mass of a point with weight $k-2$ at $G(k-2)$ and one with weight two at the origin. Therefore, $G(k)$ is on the segment $OG(2)$.

We can show similarly that for even $k \geq 3$, $G(k)$ is on the segment $OG(3)$. The proof is thus completed.   □

Theorem 1. *The strategy $\overline{\theta}$ is optimal for the basic problem. Its competitive difference is $\frac{1}{5}$.*

3.2   Structure of Optimal Offline Strategies

In the proof of Lemma 1, we have claimed that for $k \geq 2$, the strategy $(0, \frac{2\pi}{k}, \frac{4\pi}{k}, \ldots, \frac{2(k-1)\pi}{k})$ achieves a cost of zero. What should be remarked upon here that this is one of optimal offline strategies. A natural question here would be: *What other strategy achieves $C_{opt}(k) = 0$?* In what follows, we do not distinguish strategies with reflection and/or inversion symmetry or those having the same set of angles.

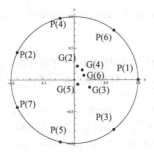 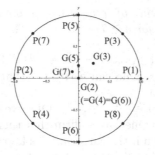 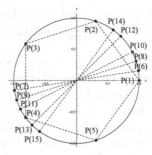

Fig. 2. Behavior of the online strategy in Theorem 2 for the 7-cyclotomic problem.

Fig. 3. Behavior of the online strategy in Theorem 3 for the 8-cyclotomic problem.

Fig. 4. Behavior of the offline strategy in Theorem 4 for the 40-cyclotomic problem, applying $p = 5$. This figure depicts the placement of 15 points such that their center of mass lies at the origin.

For $k = 2$, there does not exist such a strategy except for the strategy $(0, \pi)$. For $k = 3$, it is seen that $(0, \frac{2\pi}{3}, \frac{4\pi}{3})$ is a unique optimal offline strategy. For $k = 4$, by a basic manipulation of equations, it is derived that any optimal strategy has the form $(0, \pi, \theta_3, \theta_3 + \pi)$. Geometrically speaking, $C_\theta(4) = 0$ if and only if $P_\theta(1)P_\theta(2)P_\theta(3)P_\theta(4)$ forms a rectangle.

What if $k = 5$? Apparently, the strategy $(0, \frac{2\pi}{5}, \frac{4\pi}{5}, \frac{6\pi}{5}, \frac{8\pi}{5})$, which forms a regular pentagon, is optimal. We also have $(0, \pi, \theta_3, \theta_3 + \frac{2\pi}{3}, \theta_3 + \frac{4\pi}{3})$, for which the points compose a diameter and a regular triangle. Note that since the center of mass of a diameter and that of a regular triangle lie both at the origin, the center of mass of the five points lies at the origin as well. Then, is there any strategy that satisfies $C_\theta(5) = 0$ but does not form either a regular pentagon or the combination of a diameter and a regular triangle? The answer is yes. For $\frac{\pi}{5} > \varepsilon > 0$, we can choose $\delta > 0$ such that the strategy $(0, \frac{2\pi}{5} - \delta, \frac{4\pi}{5} + \varepsilon, \frac{6\pi}{5} - \varepsilon, \frac{8\pi}{5} + \delta)$ has a cost of zero.

4 n-Cyclotomic Problem

4.1   Simple Online Strategy

For each of the cases of odd n and even n, we provide a simple strategy and analyze its competitive difference. We will later explain that our strategy is *not* optimal in general. See Figs. 2 and 3 for the behavior.

Theorem 2. *For the n-cyclotomic problem with odd $n (\geq 3)$, the strategy $\boldsymbol{\theta}$ defined as*

$$\theta_j = \frac{(j-1)(n-1)\pi}{n}, \quad 1 \leq j \leq n$$

achieves a competitive difference of zero for $n = 3$ and $n = 5$, and a competitive difference of $\frac{1}{3\cos\frac{\pi}{14}} (< 0.35)$ for $n \geq 7$.

Proof. For $n = 3$, our strategy is $\boldsymbol{\theta} = (0, \frac{2\pi}{3}, \frac{4\pi}{3})$. By rotational symmetry, there is no choice of strategy. One can easily see that our strategy achieves a competitive difference of zero.

For $n = 5$, we have $\boldsymbol{\theta} = (0, \frac{2\pi}{5}, \frac{4\pi}{5}, \frac{6\pi}{5}, \frac{8\pi}{5})$. Observe each time when the k-th item has arrived ($1 \leq k \leq 5$). One can confirm that there is no better placement than that of our strategy, even if the cardinality is known in advance. Thus, our strategy achieves a competitive difference of zero for $n = 5$.

In the rest of the proof we discuss $n \geq 7$. We calculate the coordinate of the center of mass using the identities

$$\sum_{j=1}^{k} \cos \frac{(j-1)(n-1)\pi}{n} = \frac{1}{2\cos\frac{\pi}{2n}} \left(\cos\frac{\pi}{2n} + \sin\frac{(2k-1)(n-1)\pi)}{2n} \right),$$

$$\sum_{j=1}^{k} \sin \frac{(j-1)(n-1)\pi}{n} = \frac{1}{2\cos\frac{\pi}{2n}} \left(\sin\frac{\pi}{2n} - \sin\frac{(2k(n-1)+1)\pi}{2n} \right).$$

After some manipulation, we have

$$C_{\boldsymbol{\theta}}(k) = \frac{\left| \sin\frac{k(n-1)\pi}{2n} \right|}{k\cos\frac{\pi}{2n}}.$$

We investigate the value of $C_{\boldsymbol{\theta}}(k) - C_{opt}(k)$ for all k. For $k = 1$, we have $C_{\boldsymbol{\theta}}(1) = 1$ and obviously $C_{opt}(1) = 1$. Therefore the difference is zero. For $k = 2$, we immediately have $C_{\boldsymbol{\theta}}(2) = \sin\frac{\pi}{2n}$. By a simple calculation, it turns out that to place two points at $(1, 0)$ and $(\cos\frac{(n-1)\pi}{n}, \sin\frac{(n-1)\pi}{n})$ is optimal. Thus, $C_{opt}(2) = \sin\frac{\pi}{2n}$. Hence, the difference is again zero. For $k \geq 3$, by applying $C_{opt}(k) \geq 0$, we have

$$C_{\boldsymbol{\theta}}(k) - C_{opt}(k) \leq C_{\boldsymbol{\theta}}(k) = \frac{\left| \sin\frac{k(n-1)\pi}{2n} \right|}{k\cos\frac{\pi}{2n}}.$$

We derive

$$\frac{\left| \sin\frac{k(n-1)\pi}{2n} \right|}{k\cos\frac{\pi}{2n}} \leq \frac{1}{k\cos\frac{\pi}{2n}} \leq \frac{1}{3\cos\frac{\pi}{2n}} \leq \frac{1}{3\cos\frac{\pi}{14}},$$

since $\sin x \leq 1$ for all x, $\cos\frac{\pi}{2n}$ decreases monotonically with n, and $n \geq 7$. The competitive difference of the strategy $\boldsymbol{\theta}$ is thus upper-bounded by $\frac{1}{3\cos\frac{\pi}{14}} (< 0.35)$ for $n \geq 7$. □

We leave some remarks without proof: For $n = 7$, the strategy $(0, \frac{6\pi}{7}, \frac{10\pi}{7}, \frac{4\pi}{7}, \frac{12\pi}{7}, \frac{2\pi}{7}, \frac{8\pi}{7})$ has a competitive difference of zero, while the competitive ratio of our strategy $(0, \frac{6\pi}{7}, \frac{12\pi}{7}, \frac{4\pi}{7}, \frac{10\pi}{7}, \frac{2\pi}{7}, \frac{8\pi}{7})$ is approximately 0.08. For $n = 9$, the strategy $(0, \frac{8\pi}{9}, \frac{14\pi}{9}, \frac{4\pi}{9}, \frac{12\pi}{9}, \frac{2\pi}{9}, \frac{6\pi}{9}, \frac{16\pi}{9}, \frac{10\pi}{9})$ has a better competitive difference than that of our strategy $(0, \frac{8\pi}{9}, \frac{16\pi}{9}, \frac{6\pi}{9}, \frac{14\pi}{9}, \frac{4\pi}{9}, \frac{12\pi}{9}, \frac{2\pi}{9}, \frac{10\pi}{9})$. That is to say, our strategy is not optimal for these cases.

In addition, our strategy is not optimal for large n; roughly speaking, a strategy more like that presented in Theorem 1 performs better. More specifically, one can have a better strategy by rounding each position specified in the strategy in Theorem 1 into some nearby position that is feasible for the n-cyclotomic problem, in such a way that each position does not occur more than once. Although the rounded positions for later points may be far from those in the original strategy, this does not matter. Recall that the positions for later points do not affect the competitiveness, as we discussed in Sect. 3.1.

Theorem 3. *For the n-cyclotomic problem with even $n(\geq 2)$, the strategy $\boldsymbol{\theta}$ defined as*

$$\theta_j = \begin{cases} \frac{(j-1)\pi}{n}, & j \text{ is odd;} \\ \frac{(j-2)\pi}{n} + \pi, & j \text{ is even} \end{cases}$$

achieves a competitive difference of zero for $n = 4$, and a competitive difference of $\frac{1}{3}$ for $n \geq 6$.

Proof. For $n = 4$, the strategy obviously has a competitive difference of zero; there is no choice of strategy.

For $n \geq 6$ we first derive $C_{\boldsymbol{\theta}}(k)$ in a closed form. It is observed that every two angles in $\boldsymbol{\theta}$ place two points so that they form a diameter. Therefore, for even k, the center of mass lies at the origin and $C_{\boldsymbol{\theta}}(k) = 0$. Apparently, $C_{\boldsymbol{\theta}}(1) = 1$. What remains is odd $k \geq 3$. We have already known $C_{\boldsymbol{\theta}}(k-1) = 0$ for such k. The center of mass after placing the k-th point can be considered as the center of mass of the following two weighted points: a point with weight of $k - 1$ at the origin and one with unit weight at $P_{\boldsymbol{\theta}}(k)$ on the unit circle. Hence, the center of mass of the k points divides the line segment $OP_{\boldsymbol{\theta}}(k)$ in the ratio $1 : k - 1$. Noting $\overline{OP_{\boldsymbol{\theta}}(k)} = 1$, we have $C_{\boldsymbol{\theta}}(k) = \frac{1}{1+(k-1)} = \frac{1}{k}$ for odd k.

We next check the value of $C_{\boldsymbol{\theta}}(k) - C_{opt}(k)$ for all k. For $k = 1$, we have $C_{opt}(1) = 1$ and thus the difference is zero. For odd $k \geq 3$, we obtain

$$C_{\boldsymbol{\theta}}(k) - C_{opt}(k) \leq C_{\boldsymbol{\theta}}(k) = \frac{1}{k} \leq \frac{1}{3},$$

since $C_{opt}(k) \geq 0$ holds. For even $k \geq 2$, we have

$$C_{\boldsymbol{\theta}}(k) - C_{opt}(k) \leq C_{\boldsymbol{\theta}}(k) = 0.$$

We thus conclude that the competitive difference of the strategy $\boldsymbol{\theta}$ is at most $\frac{1}{3}$. $\square$

We add without proof that not only for $n = 4$ but also for $n = 6$, 8, and 10, our strategy is an optimal strategy. The competitive difference is $\frac{1}{3}$ for $n = 6$, $\frac{\sqrt{2}-1}{3}(\approx 0.20)$ for $n = 8$, and $\frac{\sqrt{5}-1}{6}(\approx 0.21)$ for $n = 10$. For large n, however, it turns out that our strategy is not optimal by the same reason as for large odd n.

4.2   Conditions for $C_{opt}(k) = 0$

Unlike in the basic problem, in the n-cyclotomic problem $C_{opt}(k) = 0$ is not always true for $k \geq 2$. Apart from online optimization, there arises an interesting question: *Which pair (n, k) admits $C_{opt}(k) = 0$?* In this subsection we give a partial answer. We start from easy cases.

Lemma 4. *For any n, $C_{opt}(1) = 0$, $C_{opt}(n-1) = \frac{1}{n-1}$, and $C_{opt}(n) = 0$.*

Proof. $C_{opt}(1) = 0$ and $C_{opt}(n) = 0$ are trivial. We now see why $C_{opt}(n-1) = \frac{1}{n-1}$ holds. Suppose that $n-1$ points have been placed optimally (though there is no choice) and their center of mass $G(n-1)$ lies somewhere. Next, add a point at $P(n)$, which is the unique destination without a point yet. Then, needless to say, the new center of mass comes to the origin O. By considering that the mass of the $n-1$ points concentrates at $G(n-1)$, the new center can also be thought of as the position that divides the line segment $G(n-1)P(n)$ in the ratio $1 : n-1$. Since $\overline{OP(n)} = 1$, we obtain $C_{opt}(n-1) = \overline{OG(n-1)} = \frac{1}{n-1}$.   □

The next theorem gives a sufficient condition when n belongs to a class of composite numbers.

Theorem 4. *For n even and divisible by some odd number $p \geq 3$, $C_{opt}(k) = 0$ holds if k is even or $p \leq k \leq n - p$.*

Proof. Observe that if some set of placed points forms a diameter of the unit circle or a regular polygon, then the center of mass of the points lies at the origin. The idea of our proof is thus to give a strategy that places points in such a way that they can be decomposed into such sets. If k is even, we can choose $\frac{k}{2}$ pairs of positions that form $\frac{k}{2}$ distinct diameters and the proof is done.

In what follows, assume that k is odd and satisfies $p \leq k \leq n-p$. We present a strategy for such k. For ease of presentation, only the angles appearing in the strategy are described below. Although we give a series of angles with length $n - p$ in total, the strategy is constructed so that to apply only the first k angles always leads to a cost of zero. Let $m = \frac{n}{2p}$. Intuitively, our strategy first makes a regular p-gon followed by $(m - 1)p$ distinct diameters. Formally, our strategy is to: (i) Place points at

$$0, \frac{2 \cdot 2m\pi}{n}, \frac{2 \cdot 4m\pi}{n}, \dots, \frac{2 \cdot 2(p-1)m\pi}{n}.$$

(ii) Then place points, repeatedly for $j = 1, 2, \dots, p$, at

$$\frac{2((j-1)m+1)\pi}{n}, \frac{2((j-1)m+1)\pi}{n} + \pi, \frac{2((j-1)m+2)\pi}{n}, \frac{2((j-1)m+2)\pi}{n} + \pi,$$
$$\dots, \frac{2((j-1)m+m-1)\pi}{n}, \frac{2((j-1)m+m-1)\pi}{n} + \pi.$$

It is easy to see that the placement of (i) forms a regular p-gon; the difference of the angles is all $\frac{4m\pi}{n} = \frac{2\pi}{p}$. Now $k \geq p$ is assumed, the regular p-gon is always completed.

One can see that in (ii), every pair of angles taken from the head forms a diameter. Since k and p are odd, it does not occur that at the end a diameter is left uncompleted.

Besides, it is seen that in (ii), each iteration with respect to j consists of $m - 1$ distinct diameters. What remains is to claim that any angle in (ii) does not coincide with the angles in (i). Note that $\frac{2((j-1)m+l)\pi}{n} + \pi = \frac{2((j-1)m+l+pm)\pi}{n}$. For $l = 1, 2, \cdots, m-1$, both $(j-1)m + l$ and $(j-1)m + l + pm$ are indivisible by m, which implies that none of the angles in (ii) has appeared in (i).   □

See Fig. 4 for the behavior of the strategy for $n = 40$, $p = 5$, and $k = 15$. Together with Lemma 4, we have a corollary.

Corollary 1. *For n divisible by six, $C_{opt}(k) = 0$ holds if and only if $2 \leq k \leq n - 2$ or $k = n$.*

For the case that n is a prime number, we show that $C_{opt}(k)$ cannot be zero unless $k = n$ through algebraic arguments. Let

$$B := \left\{ \left(\cos\frac{2\pi}{n}, \sin\frac{2\pi}{n}\right), \left(\cos\frac{4\pi}{n}, \sin\frac{4\pi}{n}\right), \ldots, \left(\cos\frac{2(n-1)\pi}{n}, \sin\frac{2(n-1)\pi}{n}\right)\right\},$$

which is the set of the destinations other than $(1,0)$. We here regard $\mathbb{R}^2$ as the complex plane. Then, the $n-1$ positions in B stand for the roots of the equation $z^n = 1$ except $z = 1$. Letting ζ be $\cos\frac{2\pi}{n} + i\sin\frac{2\pi}{n}$, these positions are expressed as $\zeta, \zeta^2, \ldots, \zeta^{n-1}$. The following lemma is a special case of Theorem 12.13 in [8], where $\mathbb{Q}[X]$ denotes the set of polynomials of X with the coefficients being rational.

Lemma 5. *([8]) For n prime, every element in $A := \{f(\zeta) \mid f \in \mathbb{Q}[X]\}$ can be uniquely expressed in a linear combination of $\zeta, \zeta^2, \ldots, \zeta^{n-1}$ with $a_j \in \mathbb{Q}$,*

$$a_1\zeta + a_2\zeta^2 + \cdots + a_{n-1}\zeta^{n-1}.$$

Theorem 5. *For n prime, $C_{opt}(k) = 0$ holds if and only if $k = n$.*

Proof. Apply Lemma 5 with the element to be expressed being 0. Then, the lemma states that we have to set $a_1, a_2, \ldots, a_{n-1}$ to all zero. Back in the context of $\mathbb{R}^2$, this fact implies that wherever we place k points with $1 \leq k \leq n - 1$ at some positions in B, the center of mass does not come to the origin. This is based on the observation that the linear combination for $a_j \in \{0, \frac{1}{k}\}$ with $\sum_{j=1}^{n-1} a_j = 1$ represents the center of mass of the k points placed at k distinct positions in B. By rotational symmetry, even if we use the position $(1,0)$, we know that any placement of $n - 1$ or fewer points cannot let the center come to the origin.   □

As we demonstrated in the proof of Theorem 4, $C_{opt}(k) = 0$ holds if the placement of k points can be decomposed into diameters and regular polygons. We conjecture that the converse statement is also true. Note that, on the contrary, the converse statement is false for the basic problem, as we discussed the case $k = 5$ in Sect. 3.2.

5   Concluding Remarks

Many questions are left open: What if arbitrary weights are allowed? Another measure of balance? How about in $\mathbb{R}^3$ or an arbitrary metric space? (As introduced, there are numerous studies on satellite constellation design such as [3,9].) What if the destination of points is arbitrarily restricted? For the n-cyclotomic problem, can a more sophisticated strategy be designed? Can the problem in Sect. 4.2 be solved for general composite numbers? Does $C_{opt}(k) = 0$ need the placed points to be decomposed into diameters and regular polygons?

Acknowledgments. We would like to thank the participants of the 15th Enumeration Algorithm Seminar held in Ikaho, Japan, for their helpful comments. This work was supported by KAKENHI (23700014 and 23500014).

References

1. Barba, L., De Carufel, J.L., Fleischer, R., Kawamura, A., Korman, M., Okamoto, Y., Tang, Y., Tokuyama, T., Verdonschot, S., Wang, T.: Geometric weight balancing. In: Proceedings of AAAC '13 (the 6th Annual Meeting of Asian Association for Algorithms and Computation), p. 31 (2013)
2. Borodin, A., El-Yaniv, R.: Online Computation and Competitive Analysis. Cambridge University Press, Cambridge (1998)
3. de Weck, O.L., Scialom, U., Siddiqi, A.: Optimal reconfiguration of satellite constellations with the auction algorithm. Acta Astronaut. **62**(2–3), 112–130 (2008)
4. Karger, D., Lehman, E., Leighton, T., Panigrahy, R., Levine, M., Lewin, D.: Consistent hashing and random trees: distributed caching protocols for relieving hot spots on the world wide web. In: Proceedings of STOC '97, pp. 654–663 (1997)
5. Karger, D., Sherman, A., Berkheimer, A., Bogstad, B., Dhanidina, R., Iwamoto, K., Kim, B., Matkins, L., Yerushalmi, Y.: Web caching with consistent hashing. Comput. Netw. **31**(11–16), 1203–1213 (1999)
6. Kurebe, Y., Miwa, H., Ibaraki, T.: Juuryoutsuki module tsumekomino saitekika (optimization of the packing of weighted modules). In: Proceedings of the 2007 Spring National Conference of Operations Research Society of Japan, pp. 150–151 (2007)
7. Teramoto, S., Asano, T., Katoh, N., Doerr, B.: Inserting points uniformly at every instance. IEICE Trans. Inf. Syst. E89-D(8), 2348–2356 (2006)
8. Tignol, J.P.: Galois' Theory of Algebraic Equations. World Scientific Publishing Company Incorporated, Singapore (2001)
9. Ulybyshev, Yu.: Design of satellite constellations with continuous coverage on elliptic orbits of molniya type. Cosm. Res. **47**(4), 310–321 (2009)

Transformability and Reversibility of Unfoldings of Doubly-Covered Polyhedra

Jin-ichi Itoh[1] and Chie Nara[2(✉)]

[1] Faculty of Education, Kumamoto University, Kumamoto 860-8555, Japan
j-itoh@kumamoto-u.ac.jp
[2] Liberal Arts Education Center, Aso Campus, Tokai University,
Aso, Kumamoto 869-1404, Japan
cnara@ktmail.tokai-u.jp

Abstract. Let W and X be convex polyhedra in the 3-dimensional Euclidean space. If W is dissected into a finite number of pieces which can be rearranged to form X with hinges (which compose a dissection tree), W is called *transformable* to X, and if the surface of W is transformed to the interior of X except some edges of pieces, W is called *reversible* to X. Let P be a reflective space-filler in the 3-space and let P^m be a mirror image of P. In this paper, we show that any convex unfolding W of the doubly covered polyhedron $d(P)$ of P is transformable to any convex unfolding X of the doubly covered polyhedron $d(P^m)$ of P^m, where we assume that W (resp. X) includes P (resp. P^m) as a subset. Moreover if W is dissected into n non-empty pieces (where n is the number of faces of P), W is reversible to X.

1 Introduction

The famous hinged dissection problem asked if an equilateral triangle W can be dissected into a finite number of pieces that can be rearranged to form a square X with hinges. H.E. Dudeney [5] gave an answer by giving a dissection by four pieces (see Fig. 1). There is a related topic for the n-dimensional case, the so-called Hilbert's third problem: Given any two polyhedra of equal volume, is it always possible to cut the first into finitely many polyhedral pieces which can be reassembled to yield the second? If $n = 2$, the answer is affirmative, and is known as the Bolyai-Gerwein Theorem [11]; moreover, they can be reassembled with hinges [1]. On the other hand, if $n = 3$, M. Dehn [6] gave a counterexample; the pair of a cube and a regular tetrahedron of equal volume.

We study the problem of finding a family of convex polyhedra such that any pair in the family has the above-mentioned property. If a convex polyhedron

Jin-ichi Itoh—Supported by Grant-in Aid for Scientific Research (No. 23540098), JSPS.
Chie Nara—Supported by Grant-in Aid for Scientific Research (No. 23540160), JSPS.

© Springer International Publishing Switzerland 2014
J. Akiyama et al. (Eds.): JCDCGG 2013, LNCS 8845, pp. 77–86, 2014.
DOI: 10.1007/978-3-319-13287-7_7

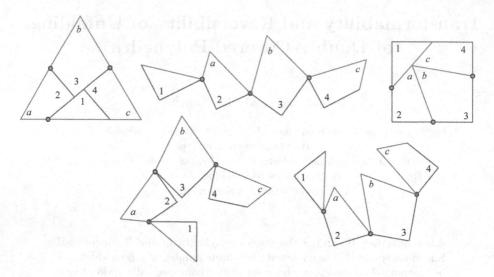

Fig. 1. Dudeney's dissection.

W in the 3-dimensional Euclidean space R^3 is dissected into a finite number of pieces which can be rearranged to form a convex polyhedron X, in such a way that there is a dissection tree whose vertex set is the set of pieces and whose edge set is the set of hinges each of which corresponds to a common edge of two pieces, we say that W is *transformable* to X. Moreover if ∂W (the surface of W) is transformed to the interior of X except some edges of pieces, we say W is *reversible* to X. J. Akiyama et al. [3] investigated reversibility of the family of canonical parallelohedra.

In this paper, we study the family of convex unfoldings of doubly-covered polyhedra. For a polyhedron P, the doubly-covered P (denoted by $d(P)$) is the degenerate 4-dimensional polytope consisting of P and its congruent copy (denoted by P^*) whose corresponding faces are identified, which means the surface of $d(P)$ is identical to that of P and the volume of an unfolding of $d(P)$ is twice that of P.

If P in R^3 is a reflective space-filling polyhedron (there are seven types of such polyhedra with no obtuse dihedral angle up to congruence and similarity [4,7]), any convex unfolding W of $d(P)$ is a *space-filler*, which means that its infinitely many congruent copies tile the space with no gaps and no 3-dimensional overlaps [8].

We call two figures in R^2 or R^3 *strictly congruent* if they can be mapped to each other by rotation and translation only (with no reflection).

We show that if P is a reflective-space filler in R^3 whose mirror image is strictly congruent to P, any two convex unfoldings W and X of $d(P)$ are transformable to each other; moreover, if W is dissected into n non-empty pieces (where n is the number of faces of P), W is reversible to X, where we assume both W and X include P as a subset. If P is a reflective-space filler in R^3 whose mirror image (denoted by P^m) is not strictly congruent to P, any unfolding W of $d(P)$ is transformable to any unfolding X of $d(P^m)$; and moreover, if W is dissected into n non-empty pieces (where n is the number of faces of P), W is reversible to X, where we assume $P \subset W$ and $P^m \subset X$.

2   Definitions and Preliminaries

Definition 1. If a convex polyhedron W is dissected into a finite number of pieces which can be rearranged to form a convex polyhedron X, in such a way that there is a dissection tree whose vertex set is the set of pieces and whose edge set is the set of hinges each of which corresponds to a common edge of two pieces, we say that W is *transformable* to X. Moreover if the surface ∂W is transformed to the interior of X except some edges of pieces, we say W is *reversible* to X. The pair of W and X is called *transformable* or *reversible*, respectively.

Note that the transformability defined in [3], is not allowed to cut any side (face) of polyhedra, but in this paper we allow to do so, because Dudeney's dissection cuts sides. Figure 2 shows the pair of a cube and a rectangular parallelepiped which is transformable, and Fig. 3 shows the pair of a rhombic dodecahedron and a rectangular parallelepiped which is reversible.

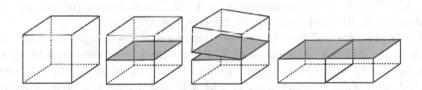

Fig. 2. The pair of a cube and a rectangular parallelepiped which is transformable.

Definition 2. A convex polyhedron W is called a *reflective space-filler* if its infinitely many congruent copies tile space (without no gaps and no 3-dim. overlaps) such that

(1) the tiling is face-to-face,
(2) if two copies have a face in common, one is obtained from the other by a reflection in the common face, and
(3) any dihedral angle of W is π/k (an integer $k \geq 2$).

The third condition is used for a simple polyhedral unfolding X of the doubly-covered W (see Definition 3) to be convex since dihedral angles of X are less than or equal twice dihedral angles of W.   H. S. M. Coxeter [4] showed that there are seven types of 3-dimensional reflective space-fillers (up to congruence and similarity): four of them are prisms (whose bases are a square, an equilateral triangle, a right triangle with an angle $\pi/3$, or an isosceles right triangle), and three of them are tetrahedra one of which has congruent triangular faces with

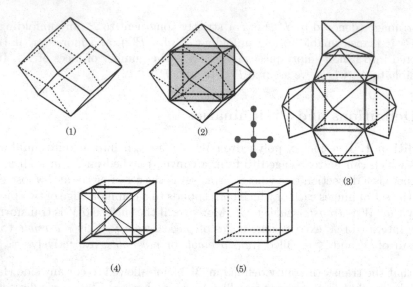

Fig. 3. The pair of a rhombic dodecahedron and a rectangular parallelepiped which is reversible.

edge lengths $\sqrt{3}$, $\sqrt{3}$ and 2, and whose dihedral angle of the edge with length 2 is $\pi/2$. We call such tetrahedron a $(\sqrt{3}, \sqrt{3}, 2)$-*tetrahedron* (or Sommerville tetrahedron, [9,10]) and denote it by ST (see Fig. 4).

By dissecting a ST into halves by the plane orthogonal to an edge with length 2, we obtain a tetrahedron Q which is also a reflective space-filler, and by dissecting Q into halves again by the plane orthogonal to the edge with length 2, we obtain a tetrahedron R which is also a reflective space-filler (see Fig. 4). We call Q and R respectively a *half-ST* and a *quarter-ST* respectively. The quarter-ST plays an important role in [2] where they call it a tetradron.

Note that the mirror image of a half-ST or a quarter-ST is *not* strictly congruent to the original.

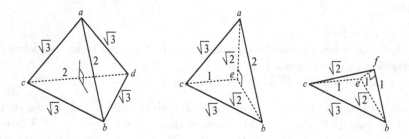

Fig. 4. Reflective space-filling tetrahedra: (1) a $(\sqrt{3}, \sqrt{3}, 2)$-tetrahedron which is denoted by ST: (2) a half-ST: (3) a quarter-ST.

Definition 3. A *doubly covered polyhedron*, denoted by $d(P)$, of a convex poly-hedron P is a degenerate 4-polytope which is composed by P and its congruent copy P^* such that corresponding faces are identified and that the surface of $d(P)$ is identical to that of P.

A body W in R^3 (which is topologically homeomorphic to a ball) is called an *unfolding* of $d(P)$ if there is a locally isometric mapping from W onto $d(P)$ whose image has no 3-dimensional overlaps. If W includes P as a subset, W is called *simple*. We call the image of the surface of W by such mapping a *cut 2-complex* of W.

For example, a rhombic dodecahedron W is an unfolding of a doubly covered cube. Dissect W into seven pieces (a cube P and six square pyramids as shown in Fig. 3(2)) and reflect each square pyramid in its square face F. The resulting figure is the doubly covered cube $d(P)$. The cut 2-complex C is a set of 12 isosceles triangles. We can also obtain $d(P)$ and C as follows: Since W is reversible to $P \cup P^m$, (where P^m is the image of P by reflection in the face F), by reflecting P^m in F we obtain $d(P)$, and C is the image of ∂W.

Figure 5 shows examples of unfoldings of $d(P)$ for a reflective space-filling tetrahedron ST. Figure 5(2) shows a cut 2-complex which is composed of six triangles, and Fig. 5(3) shows the corresponding unfolding. Figure 5(4) shows another example of an unfolding of $d(P)$, whose corresponding cut 2-complex is composed of three faces touching to the vertex a of P.

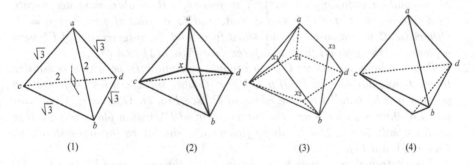

(1)   (2)   (3)   (4)

Fig. 5. (1) $d(P)$ of a tetrahedron $P = abcd$ which is a ST with edge lengths $|ab| = |cd| = 2$; (2) a cut 2-complex where x is in the interior of P^*; (3) an unfolding of $d(P)$ by the cut 2-complex showed in (2); (4) another unfolding of $d(P)$ when x in (2) is a vertex a of P.

Proposition [8]. *Any unfolding W of $d(P)$ of a reflective space-filler P is a space-filler, and its corresponding cut 2-complex C includes all edges of P (as well as P^*). If W is convex, C is composed of convex polygons, and moreover if W contains P (i.e., $C \subset P^*$), each piece of P^* which is dissected by C, has a face of P (as well as P^*), and C is characterized as follows.*

(1) If P is a rectangular parallelepiped, it is possible that C includes a rectangle (denoted by $R = abcd$) which is parallel to a face of P and whose each

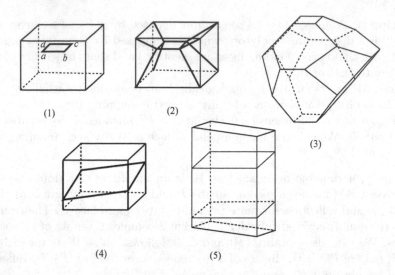

Fig. 6. (1) $d(P)$ of a rectangular parallelepiped P; (2) a cut 2-complex which includes a rectangle $abcd$ in P^*; (3) the unfolding of $d(P)$ by the cut 2-complex showed in (2); (4) another cut 2-complex of $d(P)$ when $abcd$ in (2) is a parallelogram whose vertices are on edges of P; (5) the unfolding of $d(P)$ by the cut 2-complex showed in (4).

edge is parallel to an edge of P (see Figs. 6(1) and (2)), and Fig. 6(3) is the corresponding unfolding of $d(P)$. The rectangle $R = abcd$ may degenerate to a line segment if $a = b$ and $c = d$, and to a point if $a = b = c = d$. Otherwise, C is composed of four-sided faces and the intersection of P^* with a plane intersecting all four side faces (see Figs. 6(4) and (5)).

(2) If P is a triangular prism, C may include a triangle $\triangle abc$ which is similar to the triangular face of P. If $a = b = c$, C may include a line segment (ab possible $a = b$) parallel to side edges of P (see Fig. 7). Otherwise, C is composed of three side faces and the intersection of P^* with a plane intersecting all three side faces. The resulting figures are similar to the ones shown in Figs. 6(4) and (5).

(3) If P is a tetrahedron, C may have one vertex in the interior of P^* (see Fig. 5).

In Proposition, any unfolding W of $d(P)$ is a space-filler, and the type of the tiling depends on P. For example, if P is a rectangular parallelepiped, the set of W and three congruent copies of W obtained by rotations with angle π about three edges of P, tiles the space by its translations (see [8]).

3   Theorems and Corollaries

In this section, we assume an unfolding W of $d(P)$ of a polyhedron P includes the original P (i.e., the corresponding cut 2-complex of W is contained in P^*).

Theorem 1. *Let P be a reflective space-filler whose mirror image is strictly congruent to P, that is, P be one of the following;*

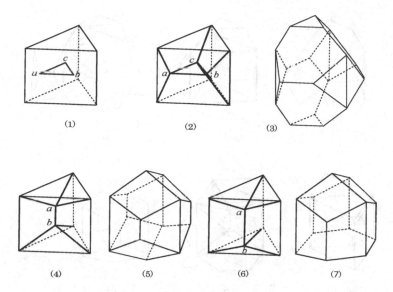

Fig. 7. (1) $d(P)$ of a prism P whose base is an equilateral triangle and a triangle similar to the base triangle abc; (2) a cut 2-complex with $\triangle abc$; (3) the unfolding of $d(P)$ by the cut 2-complex; (4) another cut 2-complex of $d(P)$ with a line segment ab parallel to side edges; (5) the unfolding of $d(P)$ corresponding to (5); (6) a cut 2-complex of $d(P)$ when b is on the face of P; (7) the unfolding of $d(P)$ corresponding to the cut 2-complex shown in (6).

(1) a parallelepiped,
(2) a triangular prism whose base is an equilateral triangle, a right triangle with
*   an angle $\pi/3$, or a isosceles right triangle, or*
(3) a $(\sqrt{3}, \sqrt{3}, 2)$-tetrahedron ST.

    Any two convex unfoldings W and X of $d(P)$ are transformable to each other, where we assume $P \subset W$ and $P \subset X$. If W divides P^* into n (non-empty) pieces by its corresponding cut 2-complex (where n is the number of edges of P), W is reversible to X (Fig. 8).

Proof. Let P^m be the image of P by reflection in the plane Π including a face of a reflective space-filler P. Then $P \cup P^m$ is an unfolding of $d(P)$ which includes P. Any unfolding W of $d(P)$ satisfying $P \subset W$, is transformable to $P \cup P^m$ by hinges $\mathcal{H}$, which is proved by a similar process to the one shown in the pair of a rhombic dodecahedron and a rectangular parallelepiped (see Fig. 3). Notice that the surface ∂W is transformed to a subset Y in P^m and Y is mapped to the cut 2-complex for W in P^* by reflection in Π.

    Let X be a convex unfolding of $d(P)$ with $P^m \subset X$ (we assume $P^m \subset X$ instead of $P \subset X$ since P^m is strictly congruent to P). Dissect P in W by the cut 2-complex which is strictly congruent to the cut 2-complex for X, and use the hinges $\mathcal{H}$. Then W is transformed to X.

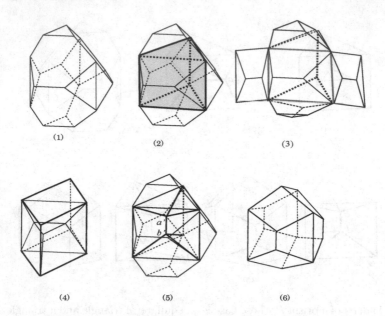

Fig. 8. (1) An unfolding W of $d(P)$ of a prism P whose base is an equilateral triangle; (2) P included in the unfolding; (3) dissection and hinges; (4) the resulting figure transformed by dissection and hinges shown in (3); (5) another dissection of W by the 2-complex strictly congruent to the mirror image of the cut 2-complex in Fig. 7(4); (6) the resulting figure by dissection and hinges shown in (5) which is strictly congruent to the polyhedron shown in Fig. 7(5).

If P^* is dissected into n (where n is the number of faces of P) pieces to obtain W, the cut 2-complex except its boundary is included in the interior of P^*, and hence ∂W is transformed to the interior of P^m except some line segments. W is reversible to X by $P^m \subset X$.

Theorem 2. *Let P be a tetrahedron which is a half-ST or a quarter-ST. Any convex unfoldings W of $d(P)$ is transformable to any convex unfolding X of $d(P^m)$, where we assume $P \subset W$ and $P^m \subset X$. If W divides P^* into four pieces by its corresponding cut 2-complex, W is reversible to X.*

Proof. The proof is similar to the one of Theorem 1, so we omit it.

If a polyhedron W is reversible to itself, W is called *self-reversible*.

Corollary 1. *Let P be a reflective space-filling polyhedron whose mirror image is strictly congruent to P, and n be the number of faces of P. If the cut 2-complex of a convex unfolding W of $d(P)$ divides P^* into n non-empty pieces, W is self-reversible, where we assume $P \subset W$.*

Proof. If an unfolding W is combinatorially equivalent to a truncated octahedron, rhombic dodecahedron, or an elongated dodecahedron, the corresponding cut 2-complex divides P^* into six non-empty pieces. So the statement in Corollary 1 follows from Theorem 1.

If a polygon T in the 2-dimensional plane is reversible to itself, T is called *self-reversible*. Note that the mirror image of a right triangle T with angle $\pi/3$ in a given plane is not strictly congruent to T in the given plane. If we restrict unfoldings of $d(P)$ in Theorem 1 to prisms, we get the following.

Corollary 2. *Let P be a polygon which is a rectangular parallelogram, an equilateral triangle, or an isosceles right triangle. Any two convex unfoldings T and S of $d(P)$ are transformable to each other, where we assume that $P \subset T$ and $P \subset S$. If T divides P^* into n pieces by its corresponding cut 2-complex (where n is the number of edges of P), T is reversible to S.*

If P is a right triangle with angles $\pi/3$, any convex unfoldings T of $d(P)$ is transformable to any unfolding S of $d(P^m)$, where we assume $P \subset T$ and $P^m \subset S$. If T divides P^* into three pieces by its corresponding cut 2-complex, T is reversible to S.

Proof. For rectangular parallelepipeds or triangular prisms P which are reflective space-fillers, consider the subfamily of simple unfoldings of $d(P)$ whose cut 2-complex is included in faces of P (as well as P^*), that is, which are orthogonal prisms. Then their images by orthogonal projection to corresponding bases are the family of unfoldings of $d(P)$.

Remark. By observing the proof of Theorem 2, we notice that the result may be extended to $d(P)$ of any convex polyhedron P whose dihedral angles are less than or equal to $\pi/2$. To do so, we only need to give precise continuous motions without self-intersection, which looks obvious, but we leave it for a future work.

References

1. Abbott, T.G., Abel, Z., Charlton, D., Demaine, E.D., Demaine, M.L., Kominers, S.D.: Hinged dissections exist. Discrete Comput. Geom. **47**(1), 150–186 (2012)
2. Akiyama, J., Kobayashi, M., Nakagawa, H., Sato, I.: Atoms for parallelohedra. Geometry intuitive, discrete and convex. Bolyai Soc. Math. Stud. **24**, 23–43 (2013). Springer
3. Akiyama, J., Sato, I., Seong, H.: On Reversibility among parallelohedra. In: Márquez, A., Ramos, P., Urrutia, J. (eds.) EGC 2011. LNCS, vol. 7579, pp. 14–28. Springer, Heidelberg (2012)
4. Coxeter, H.S.: Discrete groups generated by reflections. Ann. Math. **35**(3), 588–621 (1934)
5. Dudeney, H.E.: The Canterbury Puzzles and Other Curious Problems. W. Heinemann, London (2010)
6. Dehn, M.: Über den Rauminhalt. Math. Ann. **55**, 465–478 (1902)
7. Itoh, J., Nara, C.: Reflective space-filling polyhedra. Int. J. Pure Appl. Math. **58**(1), 87–98 (2010)
8. Itoh, J., Nara, C.: Unfoldings of doubly covered polyhedra and applications to space-fillers. Periodica Math. Hung. **63**(1), 47–64 (2011)

9. Sommerville, D.M.Y.: Space-filling tetrahedra in euclidean space. Proc. Edinburgh Math. Soc. **41**, 49–57 (1923)

10. Sommerville, D.M.Y.: Division of space by congruent triangles and tetrahedra. Proc. Edinburgh Roy. Soc. **43**, 85–116 (1923)

11. Wallace, W.: Elements of Geometry. Bell & Bradfute, Edinburgh (1831)

Computational Complexity of the r-visibility Guard Set Problem for Polyominoes

Chuzo Iwamoto$^{(\boxtimes)}$ and Toshihiko Kume

Graduate School of Engineering, Hiroshima University,
Higashi-Hiroshima 739-8527, Japan
chuzo@hiroshima-u.ac.jp

Abstract. We study the art gallery problem when the instance is a polyomino, which is the union of connected unit squares. It is shown that locating the minimum number of guards with r-visibility in a polyomino with holes is NP-hard. Here, two points u and v on a polyomino are r-visible if the orthogonal bounding rectangle for u and v lies entirely within the polyomino. As a corollary, locating the minimum number of guards with r-visibility in an orthogonal polygon with holes is NP-hard.

Keywords: Art gallery problem · Polyomino · r-visibility · NP-hard

1 Introduction

The art gallery problem is to determine the minimum number of guards who can observe the interior of a gallery. Chvátal [4] proved that $\lfloor n/3 \rfloor$ guards are the lower and upper bounds for this problem; namely, $\lfloor n/3 \rfloor$ guards are always sufficient and sometimes necessary for observing the interior of an n-vertex simple polygon. This $\lfloor n/3 \rfloor$-bound is replaced by $\lfloor n/4 \rfloor$ if the instance is restricted to a simple orthogonal polygon [7].

Another approach to the art gallery problem is to study the complexity of locating the minimum number of guards in a polygon. The NP-hardness and APX-hardness of this problem were shown by Lee and Lin [9] and by Eidenbenz et al. [5], respectively. Furthermore, Schuchardt and Hecker [13] proved that this problem remains NP-hard if we restrict our attention to simple orthogonal polygons. Even guarding the vertices of a simple orthogonal polygon was shown to be NP-hard [8].

In this paper, we study the art gallery problem when the instance is a polyomino, which is the union of connected unit squares (see Fig. 1a). It is shown that locating the minimum number of guards with r-visibility in a polyomino with holes is NP-hard. Here, two points u and v on a polyomino are said to be r-visible (or u r-sees v) if the orthogonal bounding rectangle for u and v lies entirely within the polyomino (see Fig. 1b).

As a corollary of our result, locating the minimum number of guards with r-visibility in an orthogonal polygon *with holes* is NP-hard. On the other hand, it is

J. Akiyama et al. (Eds.): JCDCGG 2013, LNCS 8845, pp. 87–95, 2014.
DOI: 10.1007/978-3-319-13287-7_8

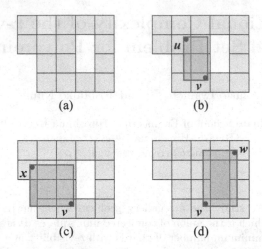

Fig. 1. (a) A polyomino P of 16 cells with a hole of two cells. (b) Two points v and u are r-visible. (c) v and x are not r-visible. (d) v and w are not r-visible.

known that the same problem for simple orthogonal polygons is polynomial-time solvable [14].

The research on the art gallery problem for polyominoes was firstly reported in [2], where it was shown that $\lfloor (m+1)/3 \rfloor$ guards are always sufficient and sometimes necessary to cover an m-polyomino (possibly with holes). Here, an m-polyomino is a connected polyomino consisting of m unit squares. Interestingly, their $\lfloor (m+1)/3 \rfloor$-bounds hold for three models of visibility: r-visibility model, all-or-nothing model, and unrestricted model (see Fig. 2).

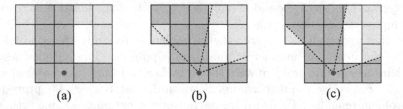

Fig. 2. (a) r-visibility model. (b) all-or-nothing model. (c) unrestricted model.

The region visible from a point in the r-visibility model is a subset of the region visible from the same point in the all-or-nothing model (see Fig. 2), which is in turn a subset of that in the unrestricted model. Due to this property, the proof of the NP-hard result for orthogonal polygons in the unrestricted model in [8] does not hold in the r-visibility model. In the conference version [1] of the above paper [2], the NP-hardness was shown for all-or-nothing and unrestricted models.

2   Definitions and Results

The definitions of polyominoes and visibility are mostly from [2,14]. A *polyomino P* is a plane geometric figure formed by joining one or more identical squares edge to edge. A polyomino is a special case of a polyform, which is a plane figure constructed by joining together identical basic polygons. We refer to the unit squares as *cells*. Figure 1(a) is an example of a polyomino P of 16 cells with a hole of two cells.

Two points v and u in P are said to be *r-visible* (or v *r-sees* u) if the orthogonal bounding rectangle for v and u lies entirely within the polyomino (see Fig. 1). Here, the rectangle may contain points on the boundary of P, but the rectangle must not contain any hole of the polyomino. (The term rectangle is used to denote the union of the boundary and of the interior.)

It is often useful to extend the notion of visibility to other geometric objects besides points. We say that two geometric objects X and Y are r-visible if and only if for all points $x \in X$ and $y \in Y$ we have that x r-sees y. For example, a cell X of a polyomino is said to be *r-visible from a point y* if every point on and inside the boundary of X is r-visible from y.

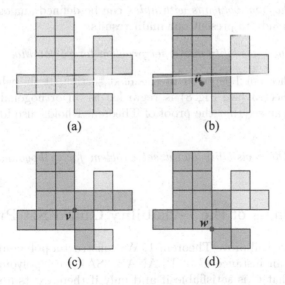

Fig. 3. The number of r-visible cells depends on the position of the point in a cell. (a) A polyomino P of 16 cells. (b) 11 cells are r-visible from point u. (c) 14 cells are r-visible from point v. (d) 13 cells are r-visible from point w.

A set of points is said to *cover* a set of cells if each of the cells is r-visible from at least one of the points. In general, the number of coverable cells from a point depends on the position of the point in the cell (see Fig. 3). However, the polyomino constructed in Sect. 3 does not depend on such positions. In other words, even if guards are placed on boundaries of cells of Fig. 8, the number of coverable cells does not change.

The definitions of a polygon and a polygon with holes are mostly from [10, 12]. A *polygon* is defined by a finite set of segments such that every segment extreme is shared by exactly two edges and no subset of edges has the same property. The segments are the *edges* and their endpoints are the *vertices* of the polygon.

If each edge of a polygon is perpendicular to one of the coordinate axes, then the polygon is called *orthogonal* or *rectilinear*. If no non-consecutive pair of edges overlap, then the polygon is said to be *simple*.

A *polygon with holes* is a polygon P enclosing several other polygons H_1, $H_2, \ldots, H_h$, the holes. None of the boundaries of P, $H_1, H_2, \ldots, H_h$ may intersect, and each of the holes is empty. P is said to bound a *multiply-connected* region with h holes: the region of the plane interior to or on the boundary of P, but exterior to or on the boundary of $H_1, H_2, \ldots, H_h$. Similarly, we define an *orthogonal polygon with holes* to be an orthogonal polygon with orthogonal holes, with all edges aligned with the same pair of orthogonal axes.

An instance of the *r-visibility guard set problem for polyominoes* is a pair (P, k), where P is a polyomino and k is a positive integer. The problem asks whether there exists a set of k points in P which covers all cells of P. The same problem *for orthogonal polygons with holes* can be defined analogously.

Now we are ready to present our main results.

Theorem 1. *The r-visibility guard set problem for polyominoes is NP-hard.*

The proof of Theorem 1 is given in the next section. If the whole polyomino constructed in Sect. 3 (see Fig. 8) is regarded as an orthogonal polygon with holes, then one can see that the proof of Theorem 1 holds also for the following corollary.

Corollary 1. *The r-visibility guard set problem for orthogonal polygons with holes is NP-hard.*

3   NP-hardness of the r-visibility Guard Set Problem

In this section, we will prove Theorem 1. We construct a polynomial-time transformation from an instance C of PLANAR 3SAT to a polyomino P and an integer k such that C is satisfiable if and only if there exists a set of k points in P covering all cells of P.

3.1   PLANAR 3SAT

The definition of PLANAR 3SAT is mostly from [LO1] on page 259 of [6]. Let $U = \{x_1, x_2, \ldots, x_n\}$ be a set of Boolean *variables*. Boolean variables take on values 0 (false) and 1 (true). If x is a variable in U, then x and $\overline{x}$ are *literals* over U. The value of $\overline{x}$ is 1 (true) if and only if x is 0 (false). A *clause* over U is a set of literals over U, such as $\{\overline{x_1}, x_3, x_4\}$. It represents the disjunction of those literals and is *satisfied* by a truth assignment if and only if at least one of its members is true under that assignment.

An instance of PLANAR 3SAT is a collection $C = \{c_1, c_2, \ldots, c_m\}$ of clauses over U such that (i) $|c_j| = 3$ or $|c_j| = 2$ for each $c_j \in C$ and (ii) the bipartite graph $B = (V, E)$, where $V = U \cup C$ and E contains exactly those pairs $\{x, c\}$ such that either literal x or $\overline{x}$ belongs to the clause c, is planar.

The PLANAR 3SAT problem asks whether there exists some truth assignment for U that simultaneously satisfies all the clauses in C. This problem is known to be NP-hard. For example, $U = \{x_1, x_2, x_3, x_4\}$, $C = \{c_1, c_2, c_3, c_4\}$, and $c_1 = \{x_1, x_2, \overline{x_3}\}$, $c_2 = \{\overline{x_1}, \overline{x_2}, x_4\}$, $c_3 = \{\overline{x_1}, x_3, \overline{x_4}\}$, $c_4 = \{\overline{x_2}, \overline{x_3}, \overline{x_4}\}$ provide an instance of PLANAR 3SAT. For this instance, the answer is "yes", since there is a truth assignment $(x_1, x_2, x_3, x_4) = (1, 0, 1, 1)$ satisfying all clauses. It is known that PLANAR 3SAT is NP-complete even if each variable occurs exactly once positively and exactly twice negatively in C [3].

3.2 Transformation from a 3SAT-instance to a Polyomino

Each variable $x_i \in \{x_1, x_2, \ldots, x_n\}$ is transformed to a polyomino of 29 cells shown in Fig. 4(a). This polyomino is called a *variable gadget*. The cells labeled with B and C correspond to $\overline{x_i}$, while the cell labeled with A corresponds to x_i. There are four possible variant forms of this gadget; cells B and C and their adjacent cells may be connected to the opposite side (see dotted cells of Fig. 4a).

Later, one can see that (a) if two guards are placed on cells B and C, then $\overline{x_i} = 1$, and (b) if a guard is placed on cell A, then $x_i = 1$ (see Fig. 5). (In this paper, a guard placed on an arbitrary point on and inside the boundary of cell g is simply called a *guard on cell g* or a *guard g*.)

Lemma 1. *(i) There is no five-guard set covering the variable gadget. (ii) There is a six-guard set covering the variable gadget such that two of the six guards are*

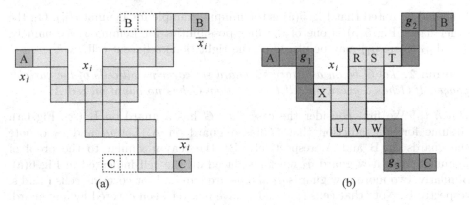

Fig. 4. (a) Variable gadget of 29 cells transformed from x_i. There are four variant forms of this gadget; cells B and C and their adjacent cells may be connected to the opposite side (see dotted cells). (b) The fourth and fifth guards are needed in R,S,T and U,V,W for covering T and W, respectively. These two guards cannot cover X.

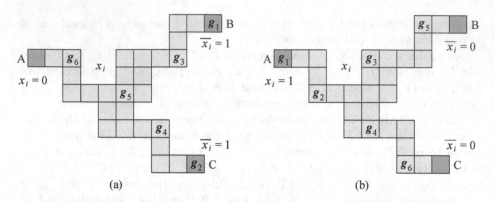

(a)   (b)

Fig. 5. Each variable gadget is covered by six guards. (a) If two guards are placed on cells B and C, then no guard is placed on A. (b) If a guard is placed on cell A, then no guard is placed on B or C.

on cells B and C. *(iii) There is a six-guard set covering the variable gadget such that one of the six guards is on A.*

Proof. (i) Consider a guard who covers cell A and can cover as many cells as possible (see g_1 in Fig. 4b). The cell on which such a guard is placed is uniquely determined. This guard g_1 can be placed on an arbitrary point on and inside the boundary of that cell. Similarly, two guards g_2 and g_3 are placed on the cells indicated in Fig. 4(b).

There are 14 cells not covered by these three guards. In order to cover cell T of Fig. 4(b), we must place a new guard on one of the cells R, S, and T. Similarly, we must place another new guard on one of the cells U, V, and W to cover W. However, these two new guards cannot cover cell X.

(ii), (iii) Such six-guard sets are given in Fig. 5(a, b), respectively.

It should be noted that Fig. 5(a) is the unique example for Lemma 1(ii). On the other hand, Fig. 5(b) is one of the four possibilities for Lemma 1(iii); namely, guard g_3 (resp. g_4) may be placed on the right (left) adjacent cell.

Lemma 2. *Let G be an arbitrary six-guard set covering all cells of the variable gadget. If G has a guard on cell B or C, then G has no guard on cell A.*

Proof. (a) We first consider the case that G has a guard on B (see Fig. 6a). Assume for contradiction that G has a guard on A. Let g_1 and g_2 denote the guards on B and A, respectively. By the reason similar to the proof of Lemma 1(i), a new guard g_3 must be placed on the cell indicated in Fig. 6(a). Similarly, two more new guards g_4 and g_5 are required for covering cells r and s, respectively. Note that cells t, u, and v have not yet been covered by any guard. Those three cells require at least two more guards, which contradicts that G is a six-guard set.

(b) The case that G has a guard on C (see Fig. 6b) is omitted, since it is very similar to (a).

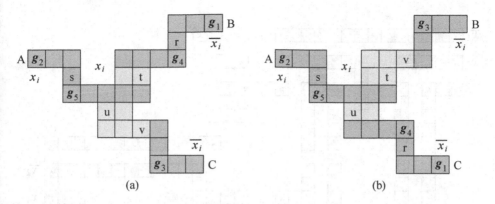

(a)   (b)

Fig. 6. There does not exist a six-guard set covering the variable gadget such that two of the guards are placed on A and B or on A and C.

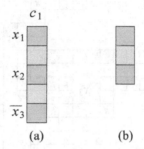

(a)   (b)

Fig. 7. (a) Clause gadget of five cells transformed from $c_1 = \{x_1, x_2, \overline{x_3}\}$. (b) If a clause consists of two literals, then the corresponding clause gadget is composed of three cells.

Each clause $c_j \in \{c_1, c_2, \ldots, c_m\}$ is transformed into a *clause gadget* of 5×1 cells if c_j has three literals (see Fig. 7a). The first, third, and fifth cells are labeled with the three literals of c_j. Those three cells are connected to the corresponding variable gadgets (see Fig. 8). If c_j consists of two literals, then the corresponding clause gadget is composed of 3×1 cells (see Fig. 7b). (The gadget of Fig. 7(b) is essential, since it is known that 3SAT WITH EXACTLY THREE OCCURRENCES PER VARIABLE is polynomial-time solvable if every clause c_j has three literals [11].)

In order to construct connections from variable gadgets to clause gadgets, we use *connection gadgets* (see green cells of Fig. 8). Each connection gadget has an even number of bends. (Bending cells are indicated by red spheres in the figure). For example, the number of bends in the connection gadget between x_4 and c_2 (resp. x_4 and c_3) is four (resp. two).

Finally, let $k = 6n + l/2 \; (= 6 \cdot 4 + 16/2 = 32)$, where n is the number of variables, and l is the number of bends in all the connection gadgets. From this construction, C is satisfiable if and only if the whole polyomino P is covered by k guards. From the positions of the 32 guards, one can see that $(x_1, x_2, x_3, x_4) = (1, 0, 1, 1)$ satisfies C. This completes the proof.

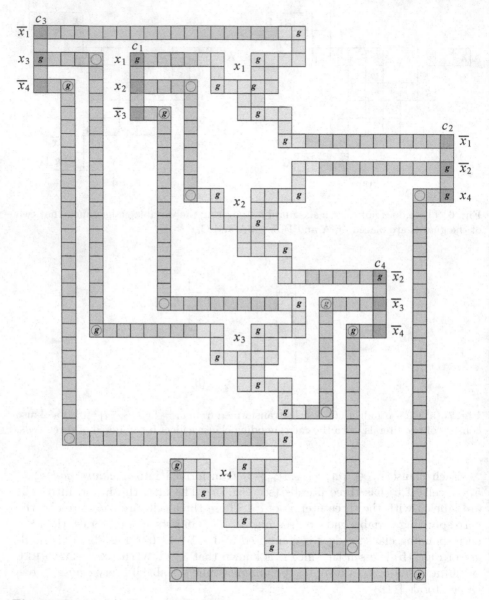

Fig. 8. The whole polyomino P transformed from $C = \{c_1, c_2, c_3, c_4\}$, where $c_1 = \{x_1, x_2, \overline{x_3}\}$, $c_2 = \{\overline{x_1}, \overline{x_2}, x_4\}$, $c_3 = \{\overline{x_1}, x_3, \overline{x_4}\}$, and $c_4 = \{\overline{x_2}, \overline{x_3}, \overline{x_4}\}$. C is satisfiable if and only if the whole polyomino is covered by $k = 32$ cells.

4   Conclusion

In this paper, we studied the art gallery problem when the instance is a polyomino with holes. It was shown that locating the minimum number of guards with r-visibility in a polyomino with holes is NP-hard. As a corollary, the same problem for orthogonal polygons with holes is NP-hard.

References

1. Biedl, T., Irfan, M.T., Iwerks, J., Kim, J., Mitchell, J.S.B.: Guarding polyominoes. In: 27th Annual Symposium on Computational Geometry, pp. 387–396 (2011)
2. Biedl, T., Irfan, M.T., Iwerks, J., Kim, J., Mitchell, J.S.B.: The art gallery theorem for polyominoes. Discrete Comput. Geom. **48**(3), 711–720 (2012)
3. Cerioli, M.R., Faria, L., Ferreira, T.O., Martinhon, C.A.J., Protti, F., Reed, B.: Partition into cliques for cubic graphs: Planar case, complexity and approximation. Discrete Appl. Math. **156**(12), 2270–2278 (2008)
4. Chvátal, V.: A combinatorial theorem in plane geometry. J. Comb. Theory B. **18**, 39–41 (1975)
5. Eidenbenz, S.J., Stamm, C., Widmayer, P.: Inapproximability results for guarding polygons and terrains. Algorithmica **31**, 79–113 (2001)
6. Garey, M.R., Johnson, D.S.: Computers and Intractability: A Guide to the Theory of NP-Completeness. W.H. Freeman, New York (1979)
7. Hoffman, F.: On the rectilinear art gallery problem. In: Paterson, M. (ed.) ICALP 1990. LNCS, vol. 443, pp. 717–728. Springer, Heidelberg (1990)
8. Katz, M.J., Roisman, G.S.: On guarding the vertices of rectilinear domains. Comp. Geom. Theor. Appl. **39**(3), 219–228 (2008)
9. Lee, D.T., Lin, A.K.: Computational complexity of art gallery problems. IEEE Trans. Inform. Theory **32**(2), 276–282 (1986)
10. O'Rourke, J.: Art Gallery Theorems and Algorithms. Oxford University Press, New York (1987)
11. Papadimitriou, C.H.: Computational Complexity. Addison-Wesley, Mass (1994)
12. Preparata, F.P., Shamos, M.I.: Computational Geometry: An Introduction. Springer, New York (1985)
13. Schuchardt, D., Hecker, H.-D.: Two NP-hard art-gallery problems for ortho-polygons. Math. Logic Quar. **41**(2), 261–267 (1995)
14. Worman, C., Keil, J.M.: Polygon decomposition and the orthogonal art gallery problem. Int. J. Comput. Geom. Ap. **17**(2), 105–138 (2007)

Properly Colored Geometric Matchings and 3-Trees Without Crossings on Multicolored Points in the Plane

Mikio Kano[1], Kazuhiro Suzuki[2]([✉]), and Miyuki Uno[3]

[1] Department of Computer and Information Sciences, Ibaraki University, Hitachi, Ibaraki, Japan
kano@mx.ibaraki.ac.jp
[2] Department of Information Science, Kochi University, Kochi, Japan
kazuhiro@tutetuti.jp
[3] University Education Center, Ibaraki University, Mito, Ibaraki, Japan
muno@mx.ibaraki.ac.jp

Abstract. Let X be a set of multicolored points in the plane such that no three points are collinear and each color appears on at most $\lceil |X|/2 \rceil$ points. We show the existence of a non-crossing properly colored geometric perfect matching on X (if $|X|$ is even), and the existence of a non-crossing properly colored geometric spanning tree with maximum degree at most 3 on X. Moreover, we show the existence of a non-crossing properly colored geometric perfect matching in the plane lattice. In order to prove these our results, we propose an useful lemma that gives a good partition of a sequence of multicolored points.

Keywords: Red and blue points · Multicolored points · Alternating matching · Alternating tree · Properly colored geometric graph · Sequence of points

MSC2010: 52C35, 05C70, 05C05.

1   Introduction

Various topics on a set of red and blue points in the plane have been studied [3]. In this paper, we consider some problems for more colors. Given a set X of multicolored points in the plane, we want to draw a graph in the plane such that the vertex set is X and each edge is a straight-line segment whose two end-vertices have distinct colors. We call such a graph a *properly colored geometric*

M. Kano—Partially supported by Japan Society for the Promotion of Science, Grant-in-Aid for Scientific Research (C).
K. Suzuki—Partially supported by MEXT. KAKENHI 24740068.
52C35: Arrangements of points, flats, hyperplanes. 05C70: Factorization, matching, partitioning, covering and packing. 05C05: Trees.

J. Akiyama et al. (Eds.): JCDCGG 2013, LNCS 8845, pp. 96–111, 2014.
DOI: 10.1007/978-3-319-13287-7_9

graph on X, which is also called an *alternating geometric graph* if X is a 2-colored point set. For alternating geometric perfect matchings on a 2-colored point set, the next theorem is well-known.

Theorem 1.1 ([5])**.** *Let R and B be sets of red and blue points in the plane, respectively. Assume that no three points of $R \cup B$ are collinear. If $|R| = |B|$, then there exists a non-crossing alternating geometric perfect matching on $R \cup B$.*

In this paper, we first generalize this Theorem 1.1 for a 3-colored point set, stated as Theorem 1.2 (Fig. 1). Note that Theorem 1.1 is a special case of Theorem 1.2 with $G = \emptyset$.

Theorem 1.2. *Let R, B, G be sets of red, blue, and green points in the plane, respectively. Assume that no three points of $X = R \cup B \cup G$ are collinear. If $|X|$ is even and each color appears on at most $|X|/2$ points, then there exists a non-crossing properly colored geometric perfect matching on X.*

Fig. 1. A non-crossing properly colored geometric perfect matching (Color figure online).

Next, we consider a tree of maximum degree at most 3, which is called a *3-tree*. Kaneko [2] proved the following theorem.

Theorem 1.3 (Kaneko [2]). *Let R and B be sets of red and blue points in the plane, respectively. Assume that no three points of $R \cup B$ are collinear. If $|R| = |B|$, then there exists a non-crossing alternating geometric spanning 3-tree on $R \cup B$.*

Our second result is a generalization of this Theorem 1.3 for a 3-colored point set, stated as Theorem 1.4 (Fig. 2).

Theorem 1.4. *Let R, B, G be sets of red, blue, and green points in the plane, respectively. Assume that no three points of $X = R \cup B \cup G$ are collinear. If each color appears on at most $\lceil |X|/2 \rceil$ points, then there exists a non-crossing properly colored geometric spanning 3-tree on X.*

If $|X|$ is even, then we can obtain this Theorem 1.4 as a corollary from our Theorem 1.2 and the following theorem by Hoffmann and Tóth [1].

Theorem 1.5 (Hoffmann and Tóth [1]). *Every disconnected properly colored geometric graph with no isolated vertices can be augmented (by adding edges) into a connected properly colored geometric graph so that the degree of every vertex increases by at most two.*

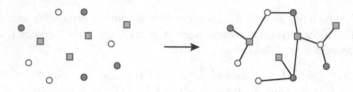

Fig. 2. A non-crossing properly colored geometric spanning 3-tree (Color figure online).

By our Theorem 1.2, there exists a non-crossing properly colored geometric perfect matching M on X. By applying Theorem 1.5 to M, we can augment M into a non-crossing properly colored geometric spanning 3-tree on X. Note that if $|X|$ is odd, then a maximum matching M on X is not a perfect matching (one isolated vertex remains), so we cannot apply Theorem 1.5 to M. In Sect. 4, we present another proof of Theorem 1.4 for both even and odd $|X|$.

We can also consider problems on red and blue points in the plane lattice by using L-line segments instead of line segments, where an *L-line segment* in the plane lattice consists of a vertical line segment and a horizontal line segment having a common endpoint. Kano et al. [4] proved the following theorem.

Theorem 1.6 (Kano and Suzuki [4]). *Let R and B be sets of red and blue points in the plane lattice, respectively. Assume that every vertical line and horizontal line passes through at most one point of the points. If $|R| = |B|$, then there exists a non-crossing alternating geometric perfect matching on $R \cup B$ such that each edge is an L-line segment.*

Our third result is a generalization of this Theorem 1.6 for a 3-colored point set, stated as Theorem 1.7 (Fig. 3).

Theorem 1.7. *Let R, B, G be sets of red, blue, and green points in the plane lattice, respectively. Assume that every vertical line and horizontal line passes through at most one point of $X = R \cup B \cup G$. If $|X|$ is even and each color appears on at most $|X|/2$ points, then there exists a non-crossing properly colored geometric perfect matching on X such that each edge is an L-line segment.*

Fig. 3. A non-crossing properly colored geometric perfect matching with L-line segments (Color figure online).

In order to prove our results, we propose the following lemma (Fig. 4).

Lemma 1.8. *Let $(x_1, x_2, \ldots, x_n)$ be a sequence of $n \geq 3$ points colored with 3 colors, say red, blue, and green. Let R, B, G be sets of red, blue, and green points in the sequence, respectively. Assume that the both ends x_1 and x_n have the same color. If each color appears on at most $\lceil n/2 \rceil$ points, then there exists an even number p $(2 \leq p \leq n - 1)$ such that x_p and x_1 have distinct colors and for every $C \in \{R, B, G\}$,*

$$|C \cap \{x_1, \ldots, x_p\}| \leq \frac{p}{2},$$

$$|C \cap \{x_{p+1}, \ldots, x_n\}| \leq \left\lceil \frac{n-p}{2} \right\rceil.$$

Fig. 4. Example of Lemma 1.8 (Color figure online).

This lemma gives a *balanced* partition of a 3-colored sequence, in the sense that every color appears on at most half of the points in each part of the partition. In our inductive proofs, this lemma is useful in some cases where some "ends" have the same color. We expect applications of the Lemma to problems where each color appears on at most half of points.

We can generalize above our results for a multicolored point set with 2, 3 or more colors, by using the following lemma.

Lemma 1.9. *Let $N_X = \{n_1, n_2, \ldots, n_r\}$ $(r \geq 4)$ be a set of positive integers. Set $n = n_1 + n_2 + \cdots + n_r$. If each integer n_i is at most $\lceil n/2 \rceil$ then there exists a tripartition $N_X = N_R \cup N_B \cup N_G$ such that*

$$\sum_{k \in N_R} k \leq \left\lceil \frac{n}{2} \right\rceil, \quad \sum_{k \in N_B} k \leq \left\lceil \frac{n}{2} \right\rceil, \quad \sum_{k \in N_G} k \leq \left\lceil \frac{n}{2} \right\rceil.$$

Proof. We may assume that $n_1 \leq n_2 \leq \cdots \leq n_r$. Then $n_1 \leq \lfloor n/r \rfloor \leq \lfloor n/4 \rfloor < \lfloor n/2 \rfloor$ since $r \geq 4$ and $n \geq 4$. Thus, for some integer n_j, it follows that $n_1 + n_2 + \cdots + n_j < \lfloor n/2 \rfloor$ and $n_1 + n_2 + \cdots + n_j + n_{j+1} \geq \lfloor n/2 \rfloor$. Then, $n_{j+2} + n_{j+3} + \cdots + n_r = n - (n_1 + n_2 + \cdots + n_j + n_{j+1}) \leq n - \lfloor n/2 \rfloor = \lceil n/2 \rceil$. Note that $1 \leq j \leq r - 2$ because if $j = r - 1$ then $n_r = n - (n_1 + \cdots + n_j) > n - \lfloor n/2 \rfloor = \lceil n/2 \rceil$, which contradicts our assumption. Hence, we have the desired tripartition $N_R = \{n_1, \ldots, n_j\}$, $N_B = \{n_{j+1}\}$, and $N_G = \{n_{j+2}, \ldots, n_r\}$. $\square$

By using Lemma 1.9, where let n_i be the number of points of color i, we can obtain the following results for multicolored points from Theorems 1.2, 1.4, 1.7, and Lemma 1.8.

Corollary 1.10. *Let X be a set of multicolored points in the plane such that no three points are collinear. If $|X|$ is even and each color appears on at most $|X|/2$ points, then there exists a non-crossing properly colored geometric perfect matching on X.*

Corollary 1.11. *Let X be a set of multicolored points in the plane such that no three points are collinear. If each color appears on at most $\lceil |X|/2 \rceil$ points, then there exists a non-crossing properly colored geometric spanning 3-trees on X.*

Corollary 1.12. *Let X be a set of multicolored points in the plane lattice. Assume that every vertical line and horizontal line passes through at most one point of X. If $|X|$ is even and each color appears on at most $|X|/2$ points, then there exists a non-crossing properly colored geometric perfect matching on X such that each edge is an L-line segment.*

Corollary 1.13. *Let $(x_1, x_2, \ldots, x_n)$ be a sequence of multicolored $n \geq 3$ points. For each color j, let C_j be a set of points colored with j in the sequence. Assume that the both ends x_1 and x_n have the same color. If $|C_j| \leq \lceil n/2 \rceil$ for every color j, then there exists an even number p $(2 \leq p \leq n-1)$ such that x_p and x_1 have distinct colors and for every color j,*

$$|C_j \cap \{x_1, \ldots, x_p\}| \leq \frac{p}{2},$$

$$|C_j \cap \{x_{p+1}, \ldots, x_n\}| \leq \left\lceil \frac{n-p}{2} \right\rceil.$$

In this paper, we will prove Lemma 1.8, Theorem 1.2, Theorem 1.4, and Theorem 1.7 in Sects. 2, 3, 4, and 5, respectively.

Throughout this paper, we will use the following definitions, notations, and a fact. For two points x and y in the plane, xy denotes the line segment joining x and y. For a set X of points in the plane, we denote by $conv(X)$ the boundary of the convex hull of X. We call a point in $X \cap conv(X)$ a *vertex* of $conv(X)$. For a graph G and its vertex v, we denote by $\deg_G(v)$ the degree of v in G. For positive integers n, a, and b such that $n = a + b$, we know that $a \leq \lceil n/2 \rceil$ and $b \leq \lceil n/2 \rceil$ if and only if $|a - b| \leq 1$. It is also known that $\lfloor n/2 \rfloor + \lceil n/2 \rceil = n$. We often use these facts without mentioning.

2   Proof of Lemma 1.8

By the symmetry of the colors, we may assume that x_1 and x_n are red. First, we claim the lemma holds when $B = \emptyset$ or $G = \emptyset$, say $G = \emptyset$.

Claim 1. *If $G = \emptyset$ then there exists an even number p $(2 \leq p \leq n-1)$ such that x_p is blue and*

$$|R \cap \{x_1, \ldots, x_p\}| = |B \cap \{x_1, \ldots, x_p\}| = \frac{p}{2},$$

$$|R \cap \{x_{p+1}, \ldots, x_n\}| \leq \left\lceil \frac{n-p}{2} \right\rceil, |B \cap \{x_{p+1}, \ldots, x_n\}| \leq \left\lceil \frac{n-p}{2} \right\rceil.$$

Proof. Define a function f from $\{1, 2, \ldots, n\}$ to the set of integers as

$$f(i) = |R \cap \{x_1, \ldots, x_i\}| - |B \cap \{x_1, \ldots, x_i\}|.$$

Then $f(i)$ increases or decreases by one, and $f(1) = |\{x_1\}| - |\emptyset| = 1$ and

$$\begin{aligned}
f(n-1) &= |R \cap \{x_1, \ldots, x_{n-1}\}| - |B \cap \{x_1, \ldots, x_{n-1}\}| = |R \setminus \{x_n\}| - |B \setminus \{x_n\}| \\
&= (|R| - 1) - |B| = |R| - 1 - (n - |R|) = 2|R| - 1 - n \\
&\leq 2\left\lceil \frac{n}{2} \right\rceil - 1 - n \leq (n+1) - 1 - n = 0.
\end{aligned}$$

Hence we can take the smallest number p $(2 \leq p \leq n - 1)$ such that $f(p) = 0$. Then, $f(p-1) = 1$. Thus, x_p is a blue point since $f(i)$ decreases when x_i is a blue point. Since $f(p) = 0$, by the definition of f, we have

$$|R \cap \{x_1, \ldots, x_p\}| = |B \cap \{x_1, \ldots, x_p\}| = \frac{p}{2}.$$

Then, p is even and for each $C \in \{R, B\}$,

$$|C \cap \{x_{p+1}, \ldots, x_n\}| = |C| - |C \cap \{x_1, \ldots, x_p\}| \leq \left\lceil \frac{n}{2} \right\rceil - \frac{p}{2} = \left\lceil \frac{n-p}{2} \right\rceil.$$

$\square$

Next, by using Claim 1, we will prove the lemma. We use induction on n. If $n = 3$ or $n = 4$ then x_i $(2 \leq i \leq n - 1)$ are not red since $x_1, x_n \in R$ and $|R| \leq \lceil n/2 \rceil = 2$. Thus, $x_p = x_2$ is the desired point. For $n \geq 5$, we suppose that the lemma holds for a sequence of $n - 2$ points.

Case 1. $|C| = \lceil n/2 \rceil$ for some $C \in \{R, B, G\}$.

Set $W = C$ and $K = (R \cup B \cup G) \setminus C$. We recolor all the points of W with white, and all the points of K with black[1]. Then, we have

$$|W| = \left\lceil \frac{n}{2} \right\rceil, |K| = n - |W| = n - \left\lceil \frac{n}{2} \right\rceil = \left\lfloor \frac{n}{2} \right\rfloor \leq \left\lceil \frac{n}{2} \right\rceil.$$

Since $x_1, x_n \in W$ or $x_1, x_n \in K$, by Claim 1, there exists an even number p $(2 \leq p \leq n - 1)$ such that x_p and x_1 have distinct colors and

$$|W \cap \{x_1, \ldots, x_p\}| = |K \cap \{x_1, \ldots, x_p\}| = \frac{p}{2},$$

$$|W \cap \{x_{p+1}, \ldots, x_n\}| \leq \left\lceil \frac{n-p}{2} \right\rceil, |K \cap \{x_{p+1}, \ldots, x_n\}| \leq \left\lceil \frac{n-p}{2} \right\rceil.$$

Hence, since each of R, B and G is a subset of W or K, the point x_p is the desired point.

---

[1] We denote a set of black points by K not by B, because B means a set of blue points in this paper.

Case 2. $|C| \leq \lceil n/2 \rceil - 1$ *for every* $C \in \{R, B, G\}$.

If x_2 is not red, then the point $x_p = x_2$ is the desired point because for every $C \in \{R, B, G\}$,

$$|C \cap \{x_3, \ldots, x_n\}| \leq |C| \leq \left\lceil \frac{n}{2} \right\rceil - 1 = \left\lceil \frac{n-2}{2} \right\rceil = \left\lceil \frac{n-p}{2} \right\rceil.$$

Hence we may assume that x_2 is red. Then there exists a blue or green point x_t $(t \geq 3)$, such that $x_1, \ldots, x_{t-1}$ are all red. We now consider a sequence

$$Y = (y_1, y_2, \ldots, y_{n-2}) = (x_2, \ldots, x_{t-1}, x_{t+1}, \ldots, x_n),$$

which is obtained from the original sequence by removing one red point x_1 and one blue or green point x_t. Note that the points $y_1 (= x_2)$ and $y_{n-2} (= x_n)$ have the same color, namely red. For every $C \in \{R, B, G\}$, $|C \cap Y| \leq |C| \leq \lceil n/2 \rceil - 1 = \lceil (n-2)/2 \rceil$. Thus, by applying the inductive hypothesis to Y, there exists an even number q $(2 \leq q \leq n-3)$ such that y_q is a blue or green point and for every $C \in \{R, B, G\}$,

$$|C \cap \{y_1, \ldots, y_q\}| \leq \frac{q}{2},$$

$$|C \cap \{y_{q+1}, \ldots, y_{n-2}\}| \leq \left\lceil \frac{n-2-q}{2} \right\rceil.$$

Then $y_q = x_{q+2}$ and $t+1 \leq q+2$ since $x_1, \ldots, x_{t-1}$ are red, $x_t \notin Y$, and y_q is not red. Hence, since x_1 and x_t have distinct colors, for every $C \in \{R, B, G\}$,

$$|C \cap \{x_1, \ldots, x_{q+2}\}| = |C \cap \{x_1, x_t, y_1, \ldots, y_q\}| \leq \frac{q}{2} + 1 = \frac{q+2}{2},$$

$$|C \cap \{x_{q+3}, \ldots, x_n\}| = |C \cap \{y_{q+1}, \ldots, y_{n-2}\}| \leq \left\lceil \frac{n-2-q}{2} \right\rceil = \left\lceil \frac{n-(q+2)}{2} \right\rceil.$$

Therefore, since q is even, namely $q+2$ is even, the point $x_p = x_{q+2}$ is the desired point.

3 Proof of Theorem 1.2 by Using Lemma 1.8

We briefly call a non-crossing properly colored geometric perfect matching a *Perfect Matching*. Set $2n = |X|$. We prove the theorem by induction on n. If $n = 1$ then the theorem is true. For $n \geq 2$, we suppose that the theorem holds for $2(n-1)$ points.

Suppose that $|C| = n$ for some $C \in \{R, B, G\}$. Set $W = C$ and $K = (R \cup B \cup G) \setminus C$. We recolor all the points of W with white, and all the points of K with black. Then there exists the desired Perfect Matching by applying Theorem 1.1 to $W \cup K$.

Hence, we may assume that

$$|C| \leq n - 1 \qquad \text{for every } C \in \{R, B, G\}.$$

Suppose that some two adjacent vertices u and v of $conv(X)$ have distinct colors. By our assumption, we have

$$|C \cap (X - \{u, v\})| \leq |C| \leq n - 1 \qquad \text{for every } C \in \{R, B, G\}.$$

Thus, since $|X - \{u, v\}| = 2(n - 1)$, we can apply the inductive hypothesis to $X - \{u, v\}$ and there exists a Perfect Matching on $X - \{u, v\}$. By adding an edge uv to this matching, we can obtain the desired Perfect Matching.

Therefore, we may assume that all the vertices of $conv(X)$ have the same color. Let v be a vertex of $conv(X)$. By a suitable rotation of the plane, we may assume that v is the highest vertex of $conv(X)$, and a and b are the left and the right vertices of $conv(X)$ adjacent to v, respectively.

We sort all the points of X with respect to their counterclockwise angle from the ray emanating from v and passing through a, and denote the sorted sequence by $(x_1, x_2, \ldots, x_{2n})$ so that $x_1 = v$, $x_2 = a$, and $x_{2n} = b$. Since the two end-points x_1 and x_{2n} have the same color, by Lemma 1.8, there exists an even number p $(2 \leq p \leq 2n - 1)$ such that for every $C \in \{R, B, G\}$,

$$|C \cap \{x_1, \ldots, x_p\}| \leq \frac{p}{2}, \quad |C \cap \{x_{p+1}, \ldots, x_{2n}\}| \leq \left\lceil \frac{2n - p}{2} \right\rceil.$$

This implies that the line passing through v and x_p partitions $X \setminus \{v, x_p\}$ into $Left = \{x_2(= a), x_3, \ldots, x_{p-1}\}$ and $Right = \{x_{p+1}, \ldots, x_{2n-1}, x_{2n}(= b)\}$ as shown in Fig. 5 so that $a \in Left$, $b \in Right$, $|Left \cup \{v, x_p\}| = p$, $|Right| = 2n - p$, and for every $C \in \{R, B, G\}$,

$$|C \cap (Left \cup \{v, x_p\})| \leq \frac{p}{2}, \quad |C \cap Right| \leq \left\lceil \frac{2n - p}{2} \right\rceil.$$

Since p is even, by applying the inductive hypothesis to each of $Left \cup \{v, x_p\}$ and $Right$, we can obtain the desired Perfect Matching.

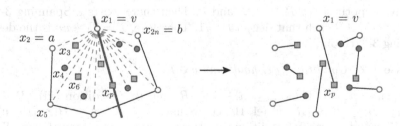

Fig. 5. A balanced partition and the desired Perfect Matching.

4   Proof of Theorem 1.4 by Using Lemma 1.8

We first prove the following proposition, which is a stronger version of Theorem 1.3. Our proof of this proposition is also another proof of Theorem 1.3.

Proposition 4.1. *Let R and B be sets of red and blue points in the plane, respectively. Assume that no three points of $X = R \cup B$ are collinear. Let v be a vertex of $conv(X)$. Suppose that one of the following three conditions (i), (ii) and (iii) holds:*

(i) $|B| = 1$, $1 \leq |R| \leq 3$, and $v \in R$,
(ii) $2 \leq |B|$, $|R| = |B| + 2$, and $v \in R$,
(iii) $2 \leq |B| \leq |R| \leq |B| + 1$.

Then there exists a non-crossing alternating geometric spanning 3-tree T on X such that $\deg_T(v) = 1$.

Proof. We briefly call an alternating geometric spanning 3-tree a *Spanning 3-Tree*. If Condition (i) holds then the star $K_{1,|R|}$, whose center is blue, is the desired Spanning 3-Tree.

Hence, we may assume that (ii) or (iii) holds. Set $n = |X|$. We prove the proposition by induction on n. By the assumption of the proposition, $n \geq 4$. If $n = 4$ then $|R| = |B| = 2$. Thus, there exists a non-crossing alternating geometric matching $M = \{va, bc\}$ where a and b have distinct colors. Then, the path $vabc$ is the desired Spanning 3-Tree.

For $n \geq 5$, we suppose that the proposition holds for at most $n-1$ points. The outline of the proof is that we will find a Spanning 3-Tree on $X - v$ and connect v and a point with degree at most 2 in the tree. We consider the following two cases depending on the colors of the two neighbors of v of $conv(X)$.

Case 1. *v and a neighbor vertex u of v of $conv(X)$ have distinct colors.*

Subcase 1.1. *Condition (ii) holds.*

Since $v \in R$ and $|R| = |B| + 2$, $X - v = (R - v) \cup B$ and $|R - v| = |B| + 1$. Since $2 \leq |B|$, we have $2 \leq |B| \leq |R - v| \leq |B| + 1$. Thus, $R - v$ and B satisfy Condition (iii). Hence, since u is a vertex of $conv(X - v)$, we can apply the inductive hypothesis to $R - v$, B, and u. Then there exists a Spanning 3-Tree T_1 on $(R - v) \cup B$ such that $\deg_{T_1}(u) = 1$. Therefore, $T = T_1 + vu$ is the desired Spanning 3-Tree on X.

Subcase 1.2. *Condition (iii) holds and $v \in R$.*

$3 \leq |R|$ since $n \geq 5$. Thus, $2 \leq |R| - 1 \leq |R - v|$. By Condition (iii), $|R - v| = |R| - 1 \leq |B| \leq |R| = |R - v| + 1$. Hence, we have $2 \leq |R - v| \leq |B| \leq |R - v| + 1$. Thus, $R - v$ and B satisfy Condition (iii). Hence, since u is a vertex of $conv(X - v)$, we can apply the inductive hypothesis to $R - v$, B, and u. Then there exists a Spanning 3-Tree T_1 on $(R - v) \cup B$ such that $\deg_{T_1}(u) = 1$. Therefore, $T = T_1 + vu$ is the desired Spanning 3-Tree on X.

Subcase 1.3. *Condition (iii) holds and $v \in B$.*

If $|B| = 2$ then $2 \le |R| \le 3$ and $|B - v| = 1$. Thus, since $u \in R$, R and $B - v$ satisfy Condition (i). If $3 \le |B|$ then $2 \le |B - v|$. By Condition (iii), we have $2 \le |B - v| \le |B| \le |R| \le |B| + 1 = |B - v| + 2$. Thus, since $u \in R$, R and $B - v$ satisfy Condition (ii) or (iii). Hence, since u is a vertex of $conv(X - v)$, we can apply the inductive hypothesis to R, $B - v$, and u. Then there exists a Spanning 3-Tree T_1 on $R \cup (B - v)$ such that $\deg_{T_1}(u) = 1$. Therefore, $T = T_1 + vu$ is the desired Spanning 3-Tree on X.

Case 2. *v and its two neighbor vertices of $conv(X)$ have the same color.*

By a suitable rotation of the plane, we may assume that v is the highest vertex of $conv(X)$, and a and b are the left and the right vertices of $conv(X)$ adjacent to v, respectively.

Subcase 2.1. *$v, a, b \in R$.*

We sort all the points of $X - v$ with respect to their counterclockwise angle from the ray emanating from v and passing through a, and denote the sorted sequence by $(x_1, x_2, \ldots, x_{n-1})$ so that $x_1 = a$ and $x_{n-1} = b$. Since $2 \le |B| \le |R| \le |B| + 2$, we have $1 \le |B| - 1 \le |R - v| \le |B| + 1$, which implies $||R - v| - |B|| \le 1$. Thus, in the sequence, each color appears on at most $\lceil (n-1)/2 \rceil$ points. Since the two end-points x_1 and x_{n-1} have the same color, namely red, by Lemma 1.8, there exists an even number p $(2 \le p \le n - 2)$ such that $x_p \in B$ and for every $C \in \{R, B\}$,

$$|C \cap \{x_1, \ldots, x_p\}| \le \frac{p}{2}, \quad |C \cap \{x_{p+1}, \ldots, x_{n-1}\}| \le \left\lceil \frac{n-1-p}{2} \right\rceil.$$

This implies that the line passing through v and x_p partitions $X \setminus \{v, x_p\}$ into $Left = \{x_1(= a), x_2, \ldots, x_{p-1}\}$ and $Right = \{x_{p+1}, \ldots, x_{n-2}, x_{n-1}(= b)\}$ as shown in Fig. 6 so that $a \in Left$, $b \in Right$, $|Left \cup \{x_p\}| = p$, $|Right| = n - 1 - p$, and for every $C \in \{R, B\}$,

$$|C \cap (Left \cup \{x_p\})| \le \frac{p}{2}, \quad |C \cap Right| \le \left\lceil \frac{n-1-p}{2} \right\rceil. \tag{1}$$

Here, we will find two Spanning 3-Trees T_1 and T_2 on $Left \cup \{x_p\}$ and $Right \cup \{x_p\}$ such that $\deg_{T_1}(x_p) = 1$ and $\deg_{T_2}(x_p) = 1$, respectively, and connect the red point v to the blue point x_p.

First, set $W = R \cap Left$ and $K = (B \cap Left) \cup \{x_p\}$. Since p is even, $|K| = |W|$. Hence, since $x_p \in K$ is a vertex of $conv(K \cup W)$, we can apply the inductive hypothesis to K, W, and x_p. Then there exists a Spanning 3-Tree T_1 on $Left \cup \{x_p\}$ such that $\deg_{T_1}(x_p) = 1$.

Next, set $W = R \cap Right$ and $K = (B \cap Right) \cup \{x_p\}$. By the inequality (1), $||W| - (|K| - 1)| = ||R \cap Right| - |B \cap Right|| \le 1$. Thus, we have $-1 \le |W| - (|K| - 1) \le 1$, that is, $0 \le |K| - |W| \le 2$. Then, either of the following

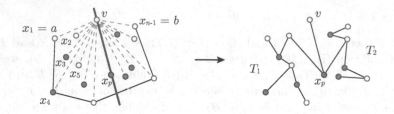

Fig. 6. A balanced partition and the desired Spanning 3-tree (Color figure online).

conditions (i), (ii), or (iii) holds: (i) $|W| = 1$, $1 \le |K| \le 3$, and $x_p \in K$, (ii) $2 \le |W|$, $|K| = |W| + 2$, and $x_p \in K$, (iii) $2 \le |W| \le |K| \le |W| + 1$. Hence, since $x_p \in K$ is a vertex of $conv(K \cup W)$, we can apply the inductive hypothesis to K, W, and x_p. Then there exists a Spanning 3-Tree T_2 on $Right \cup \{x_p\}$ such that $\deg_{T_2}(x_p) = 1$.

Consequently, $T = T_1 + T_2 + vx_p$ is the desired Spanning 3-tree on X.

Subcase 2.2. $v, a, b \in B$.

We sort all the points of X with respect to their counterclockwise angle from the ray emanating from v and passing through a, and denote the sorted sequence by $(x_1, x_2, \ldots, x_n)$ so that $x_1 = v$, $x_2 = a$, and $x_n = b$. Note that in this subcase R and B satisfy Condition (iii) since $v \in B$, that is, $2 \le |B| \le |R| \le |B| + 1$. Thus, in the sequence, each color appears on at most $\lceil n/2 \rceil$ points. Since the two end-points x_1 and x_n have the same color, namely blue, by Lemma 1.8, there exists an even number p ($2 \le p \le n - 1$) such that $x_p \in R$ and for every $C \in \{R, B\}$,

$$|C \cap \{x_1, \ldots, x_p\}| \le \frac{p}{2}, \quad |C \cap \{x_{p+1}, \ldots, x_n\}| \le \left\lceil \frac{n-p}{2} \right\rceil.$$

This implies that the line passing through v and x_p partitions $X \setminus \{v, x_p\}$ into $Left = \{x_2(= a), x_3, \ldots, x_{p-1}\}$ and $Right = \{x_{p+1}, \ldots, x_{n-1}, x_n(= b)\}$ as shown in Fig. 7 so that $a \in Left$, $b \in Right$, $|Left \cup \{x_1(= v), x_p\}| = p$, $|Right| = n - p$, and for every $C \in \{R, B\}$,

$$|C \cap (Left \cup \{v, x_p\})| \le \frac{p}{2}, \quad |C \cap Right| \le \left\lceil \frac{n-p}{2} \right\rceil. \tag{2}$$

Here, we will find two Spanning 3-Trees T_1 and T_2 on $Left \cup \{x_p\}$ and $Right \cup \{x_p\}$ such that $\deg_{T_1}(x_p) = 1$ and $\deg_{T_2}(x_p) = 1$, respectively, and connect the blue point v to the red point x_p.

First, set $W = (R \cap Left) \cup \{x_p\}$ and $K = B \cap Left$. Since $Left \cup \{x_1(= v), x_p\}$ has even points and $x_1(= v)$ is blue, $|W| = |K| + 1$. Hence, since $x_p \in W$ is a vertex of $conv(W \cup K)$, we can apply the inductive hypothesis to W, K, and x_p. Then there exists a Spanning 3-Tree T_1 on $Left \cup \{x_p\}$ such that $\deg_{T_1}(x_p) = 1$.

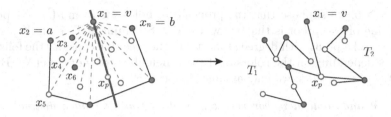

Fig. 7. A balanced partition and the desired Spanning 3-tree (Color figure online).

Next, set $W = (R \cap Right) \cup \{x_p\}$ and $K = B \cap Right$. By the inequality (2), $|(|W| - 1) - |K|| = ||R \cap Right| - |B \cap Right|| \le 1$. Thus, we have $-1 \le (|W| - 1) - |K| \le 1$, that is, $0 \le |W| - |K| \le 2$. Then, either of the following conditions (i), (ii), or (iii) holds: (i) $|K| = 1$, $1 \le |W| \le 3$, and $x_p \in W$, (ii) $2 \le |K|$, $|W| = |K| + 2$, and $x_p \in W$, (iii) $2 \le |K| \le |W| \le |K| + 1$. Hence, since $x_p \in W$ is a vertex of $conv(W \cup K)$, we can apply the inductive hypothesis to W, K, and x_p. Then there exists a Spanning 3-Tree T_2 on $Right \cup \{x_p\}$ such that $\deg_{T_2}(x_p) = 1$.

Consequently, $T = T_1 + T_2 + vx_p$ is the desired Spanning 3-tree on X. $\square$

Now, we will prove Theorem 1.4. If $|X| \le 3$ then the theorem is true. Thus, we may assume that $|X| \ge 4$. Instead of Theorem 1.4 with $|X| \ge 4$, we prove the following stronger Proposition 4.2 by using Lemma 1.8 and Proposition 4.1.

Proposition 4.2. *Let R, B, G be sets of red, blue, and green points in the plane, respectively. Assume that no three points of $X = R \cup B \cup G$ are collinear. $|X| \ge 4$. Let v be a vertex of $conv(X)$. If each color appears on at most $\lceil |X|/2 \rceil$ points, then there exists a non-crossing properly colored geometric spanning 3-tree T on X such that $\deg_T(v) = 1$.*

Proof. Set $n = |X|$. We briefly call a properly colored geometric spanning 3-tree a *Spanning 3-Tree*. If there are exactly two colors then the proposition holds by Proposition 4.1. Thus, we may assume that $R \ne \emptyset$, $B \ne \emptyset$, $G \ne \emptyset$.

Suppose that the number of points colored with some color is exactly $\lceil n/2 \rceil$, say $|R| = \lceil n/2 \rceil$. Then $|B \cup G| = n - |R| = \lfloor n/2 \rfloor \le \lceil n/2 \rceil$. Set $W = R$ and $K = B \cup G$. Then, $2 \le |K| \le |W| \le |K| + 1$ since $n \ge 4$. Thus, we can apply Proposition 4.1 to W, K, and v. Then there exists a Spanning 3-Tree T with $\deg_T(v) = 1$ on $X = W \cup K$, which is the desired tree. Therefore, we have the following claim.

Claim 1. *We may assume that each color appears on at most $\lceil n/2 \rceil - 1$ points.*

We prove Proposition 4.2 by induction on n. If $n = 4$ then we may assume that $|R| = 2$, $|B| = 1$, and $|G| = 1$ by the symmetry of the colors. Set $W = R$ and $K = B \cup G$. Then, $2 \le |W| = |K|$. Thus, we can apply Proposition 4.1 to W, K, and v. Then there exists a Spanning 3-Tree T on $X = W \cup K$ such that $\deg_T(v) = 1$, which is the desired tree.

For $n \geq 5$, we suppose that the proposition holds for at most $n - 1$ points. The outline of the proof is that we will find a Spanning 3-Tree on $X - v$ and connect v and a point with degree at most 2 in the tree. We consider the following two cases depending on the colors of the two neighbors of v of $conv(X)$. By the symmetry of the colors, we may assume that $v \in R$.

Case 1. *v and some neighbor vertex u of v of $conv(X)$ have distinct colors, namely, $u \notin R$.*

$|X - v| \geq 4$. By Claim 1, for every $C \in \{R, B, G\}$, $|C - v| \leq |C| \leq \lceil n/2 \rceil - 1 \leq \lceil |X - v|/2 \rceil$ points. Hence, since u is a vertex of $conv(X - v)$, we can apply the inductive hypothesis to $X - v$ and u. Then there exists a Spanning 3-Tree T_1 on $X - v$ such that $\deg_{T_1}(u) = 1$. Therefore, $T = T_1 + vu$ is the desired Spanning 3-Tree on X.

Case 2. *v and its two neighbor vertices of $conv(X)$ have the same color.*

By a suitable rotation of the plane, we may assume that v is the highest vertex of $conv(X)$, and a and b are the left and the right vertices of $conv(X)$ adjacent to v, respectively. We sort all the points of $X - v$ with respect to their counterclockwise angle from the ray emanating from v and passing through a, and denote the sorted sequence by $(x_1, x_2, \ldots, x_{n-1})$ so that $x_1 = a$ and $x_{n-1} = b$.

By Claim 1, for every $C \in \{R, B, G\}$, $|C - v| \leq |C| \leq \lceil n/2 \rceil - 1 \leq \lceil (n-1)/2 \rceil$ points. Thus, in the sequence, each color appears on at most $\lceil |X - v|/2 \rceil$ points. The two end-points x_1 and x_{n-1} have the same color, namely red. Hence, by Lemma 1.8, there exists an even number p ($2 \leq p \leq n - 2$) such that $x_p \notin R$ and for every $C \in \{R, B, G\}$,

$$|C \cap \{x_1, \ldots, x_p\}| \leq \frac{p}{2}, \quad |C \cap \{x_{p+1}, \ldots, x_{n-1}\}| \leq \left\lceil \frac{n-1-p}{2} \right\rceil.$$

This implies that the line passing through v and x_p partitions $X \setminus \{v, x_p\}$ into $Left = \{x_1(= a), x_2, \ldots, x_{p-1}\}$ and $Right = \{x_{p+1}, \ldots, x_{n-2}, x_{n-1}(= b)\}$ as shown in Fig. 8 so that $a \in Left$, $b \in Right$, $|Left \cup \{x_p\}| = p$, $|Right| = n - 1 - p$, and for every $C \in \{R, B, G\}$,

$$|C \cap (Left \cup \{x_p\})| \leq \frac{p}{2}, \quad |C \cap Right| \leq \left\lceil \frac{n-1-p}{2} \right\rceil. \tag{3}$$

Here, we will find two Spanning 3-Trees T_1 and T_2 on $Left \cup \{x_p\}$ and $Right \cup \{x_p\}$ such that $\deg_{T_1}(x_p) = 1$ and $\deg_{T_2}(x_p) = 1$, respectively, and connect the red point v to the non-red point x_p. By the symmetry of B and G, we may assume that $x_p \in B$.

First, we will find a Spanning 3-Tree T_1 on $Left \cup \{x_p\}$ such that $\deg_{T_1}(x_p) = 1$. If $|Left \cup \{x_p\}| = 2$ then the path $x_p a$ is the desired Spanning 3-Tree T_1. Thus, since p is even, we suppose that $|Left \cup \{x_p\}| \geq 4$. Then, since x_p is a vertex of $conv(Left \cup \{x_p\})$, we can apply the inductive hypothesis to $Left \cup \{x_p\}$ and x_p. Then there exists a Spanning 3-Tree T_1 on $Left \cup \{x_p\}$ such that $\deg_{T_1}(x_p) = 1$.

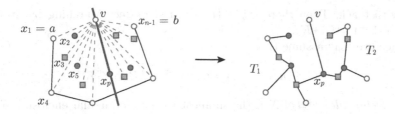

Fig. 8. A balanced partition and the desired Spanning 3-tree (Color figure online).

Next, we will find a Spanning 3-Tree T_2 on $Right \cup \{x_p\}$ such that $\deg_{T_2}(x_p) = 1$. If $|Right \cup \{x_p\}| = 2$ then the path $x_p b$ is the desired Spanning 3-Tree T_2. If $|Right \cup \{x_p\}| = 3$ then $n - 1 - p = |Right| = 2$. Thus, by the inequality (3), $|C \cap Right| \leq 1$. Thus implies that $Right \cup \{x_p\}$ has one red point b, one blue point x_p, and one blue or green point g. Hence, the path $x_p bg$ is the desired Spanning 3-Tree T_2.

Thus, we suppose that $|Right \cup \{x_p\}| \geq 4$. If for every $C \in \{R, B, G\}$, $|C \cap (Right \cup \{x_p\})| \leq \lceil (n-p)/2 \rceil$, then, since x_p is a vertex of $conv(Right \cup \{x_p\})$, we can apply the inductive hypothesis to $Right \cup \{x_p\}$ and x_p. Then there exists a Spanning 3-Tree T_2 on $Right \cup \{x_p\}$ such that $\deg_{T_2}(x_p) = 1$.

Hence, we suppose that for some $C \in \{R, B, G\}$, $|C \cap (Right \cup \{x_p\})| > \lceil (n-p)/2 \rceil$. Since x_p is blue, by the inequality (3), we have

$$\left\lceil \frac{n-p}{2} \right\rceil < |B \cap (Right \cup \{x_p\})| \leq \left\lceil \frac{n-1-p}{2} \right\rceil + 1 = \left\lceil \frac{n-p+1}{2} \right\rceil.$$

This implies that $n - p$ is even and $|B \cap (Right \cup \{x_p\})| = (n-p)/2 + 1$. Set $W = B \cap (Right \cup \{x_p\})$ and $K = (Right \cup \{x_p\}) \setminus W$. Then,

$$|K| = |Right \cup \{x_p\}| - |W| = (n - 1 - p + 1) - \left(\frac{n-p}{2} + 1\right) = \frac{n-p}{2} - 1$$

Thus, $|W| = |K| + 2$. Hence, since $x_p \in W$ is a vertex of $conv(W \cup K)$, we can apply Proposition 4.1 to W, K, and x_p. Then there exists a Spanning 3-Tree T_2 on $Right \cup \{x_p\}$ such that $\deg_{T_2}(x_p) = 1$.

Consequently, $T = T_1 + T_2 + vx_p$ is the desired Spanning 3-tree on X. □

5 Proof of Theorem 1.7 by Using Lemma 1.8

We can prove Theorem 1.7 in the same way as the proof of Theorem 1.2 in Sect. 3. We briefly call a non-crossing properly colored geometric perfect matching (such that each edge is an L-line segment) a *Perfect Matching*. Set $2n = |X|$. We prove the theorem by induction on n. If $n = 1$ then the theorem is true. For $n \geq 2$, we suppose that the theorem holds for $2(n - 1)$ points.

Suppose that $|C| = n$ for some $C \in \{R, B, G\}$. Set $W = C$ and $K = (R \cup B \cup G) \setminus C$. We recolor all the points of W with white, and all the points of K with black. Then there exists the desired Perfect Matching by applying Theorem 1.6 to $W \cup K$.

Hence, we may assume that

$$|C| \le n - 1 \qquad \text{for every } C \in \{R, B, G\}.$$

The *rectangular hull* of X is the smallest closed rectangular enclosing X. We denote by $rect(X)$ the boundary of the rectangular hull of X. We call a point in $X \cap rect(X)$ a *vertex* on $rect(X)$.

Suppose that some two adjacent vertices u and v on $rect(X)$ have distinct colors. By our assumption, we have

$$|C \cap (X - \{u, v\})| \le |C| \le n - 1 \qquad \text{for every } C \in \{R, B, G\}.$$

Thus, since $|X - \{u, v\}| = 2(n - 1)$, we can apply the inductive hypothesis to $X - \{u, v\}$ and there exists a Perfect Matching on $X - \{u, v\}$. By adding an L-line segment uv on $rect(X)$ to this matching, we can obtain the desired Perfect Matching.

Therefore, we may assume that all the vertices on $rect(X)$ have the same color. Let a and b be the left and the right vertices on $rect(X)$, respectively. We sort all the points of X by their horizontal coordinate, and denote the sorted sequence by $(x_1, x_2, \ldots, x_{2n})$ so that $x_1 = a$ and $x_{2n} = b$. Since the two endpoints x_1 and x_{2n} have the same color, by Lemma 1.8, there exists an even number p $(2 \le p \le 2n - 1)$ such that for every $C \in \{R, B, G\}$,

$$|C \cap \{x_1, \ldots, x_p\}| \le \frac{p}{2}, \quad |C \cap \{x_{p+1}, \ldots, x_{2n}\}| \le \left\lceil \frac{2n - p}{2} \right\rceil.$$

This implies that the vertical line passing through x_p partitions $X \setminus \{x_p\}$ into $Left = \{x_1(= a), x_2, \ldots, x_{p-1}\}$ and $Right = \{x_{p+1}, \ldots, x_{2n-1}, x_{2n}(= b)\}$ so that $a \in Left$, $b \in Right$, $|Left \cup \{x_p\}| = p$, $|Right| = 2n - p$, and for every $C \in \{R, B, G\}$,

$$|C \cap (Left \cup \{x_p\})| \le \frac{p}{2}, \quad |C \cap Right| \le \left\lceil \frac{2n - p}{2} \right\rceil.$$

Since p is even, by applying the inductive hypothesis to each of $Left \cup \{x_p\}$ and $Right$, we can obtain the desired Perfect Matching.

References

1. Hoffmann, M., Tóth, C.D.: Vertex-colored encompassing graphs. Graphs and Combinatorics **30**, 933–947 (2014)
2. Kaneko, A.: On the maximum degree of bipartite embeddings of trees in the plane. In: Akiyama, J., Kano, M., Urabe, M. (eds.) JCDCG 1998. LNCS, vol. 1763, pp. 166–171. Springer, Heidelberg (2000)

3. Kaneko, A., Kano, M.: Discrete geometry on red and blue points in the plane — a survey —. In: Aronov, B., Basu, S., Pach, J., Sharir, M. (eds.) Discrete and Computational Geometry. Algorithms and Combinatorics, vol. 25, pp. 551–570. Springer, Heidelberg (2003)
4. Kano, M., Suzuki, K.: Discrete geometry on red and blue points in the plane lattice. In: Pach, J. (ed.) Thirty Essays on Geometric Graph Theory, pp. 355–369. Springer, New York (2013)
5. Larson, L.C.: Problem-Solving Through Problems. Problem Books in Mathematics. Springer, New York (1983)

Generating Polygons with Triangles

T. Kuwata[1](✉) and H. Maehara[2]

[1] Tokai University, Kanagawa, Japan
kuwata@tokai-u.jp
[2] Ryukyu University, Okinawa, Japan

Abstract. A set of triangles $\mathcal{F}$ is said to generate a polygon P if a homothetic transform λP of P can be dissected into triangles each congruent to a triangle in $\mathcal{F}$. The simplicial element number of a polygon P is defined to be the minimum cardinality of a family $\mathcal{F}$ of triangles that can generate P. The simplicial element number of a set of polygons $P_1, P_2, \ldots, P_k$ is defined to be the minimum cardinality of a family $\mathcal{F}$ of triangles that can generate all $P_1, \ldots, P_k$. In this paper, we consider simplicial element numbers for several set of regular polygons and generating relations among triangles.

Keywords: Original triangle · Terminal triangle · Intermediate triangle · Simplicial element number · Generating chain

Mathematical subject classification (2010): 52B45 · 52C20

1 Introduction

A finite set of simplices $\mathcal{F}$ is said to *tile* a polytope P, if P can be represented as the union of a number of simplices $\tau_1, \tau_2, \ldots, \tau_N$ such that the interiors of these τ_i are mutually disjoint, and each τ_i is congruent to a member of $\mathcal{F}$. A set of simplices $\mathcal{F}$ is said to *generate* a polytope P, written as $\mathcal{F} \to P$, if $\mathcal{F}$ tiles a polytope that is similar to P. For a simplex σ, we write $\sigma \to P$ instead of $\{\sigma\} \to P$. Note that $(\sigma \to \tau) \wedge (\tau \to P)$ implies $\sigma \to P$.

The *simplicial element number* of a polytope P, denoted by $e(P)$, is defined to be the minimum cardinality of a family $\mathcal{F}$ of simplices that can generate P. The simplicial element number of a set of polytopes $P_1, P_2, \ldots, P_k$, denoted by $e(P_1, P_2, \ldots, P_k)$, is defined to be the minimum cardinality of a family $\mathcal{F}$ of simplices that can generate all $P_1, \ldots, P_k$. For example, $e(\{3\}, \{6\}) = 1$, where $\{p\}$ denotes the regular p-gon. It is obvious that the inequality $e(P_1, \ldots, P_k) \leq e(P_1) + \cdots + e(P_k)$ holds. If equality holds, then $P_1, \ldots, P_k$ are said to be *independent*.

We denote $\sigma \sim \tau$ if two simplices σ and τ are similar. For a simplex σ, define as follows:

$$\sigma \text{ is} \begin{cases} original & \text{if } \tau \to \sigma \text{ implies } \tau \sim \sigma, \\ terminal & \text{if } \sigma \to \tau \text{ implies } \tau \sim \sigma, \\ intermediate & \text{otherwise.} \end{cases}$$

© Springer International Publishing Switzerland 2014
J. Akiyama et al. (Eds.): JCDCGG 2013, LNCS 8845, pp. 112–121, 2014.
DOI: 10.1007/978-3-319-13287-7_10

A finite sequence of simplices $\sigma_1\sigma_2\dots\sigma_k$ (or written as $\sigma_1 \to \sigma_2 \to \cdots \to \sigma_k$) is called a *generating chain of length* k if $\sigma_i \to \sigma_{i+1}$ and $\sigma_i \not\sim \sigma_{i+1}$ for $i = 1, 2, \dots, k-1$. For example, in the two-dimensional case,

$$\triangle(\tfrac{\pi}{6}, \tfrac{\pi}{3}, \tfrac{\pi}{2}) \to \triangle(\tfrac{\pi}{6}, \tfrac{\pi}{6}, \tfrac{2\pi}{3}) \to \triangle(\tfrac{\pi}{3}, \tfrac{\pi}{3}, \tfrac{\pi}{3}) \qquad (*)$$

is a generating chain of length 3 (see Fig. 1), where $\triangle(\alpha, \beta, \gamma)$ denotes a triangle with interior angles α, β, γ. Note that, as far as generating relations concern, we need not distinguish two similar triangles. So, we may regard $\triangle(\alpha, \beta, \gamma)$ as any representative of the triangles with interior angles α, β, γ.

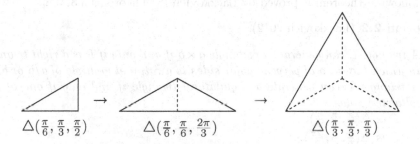

$$\triangle(\tfrac{\pi}{6}, \tfrac{\pi}{3}, \tfrac{\pi}{2}) \qquad \triangle(\tfrac{\pi}{6}, \tfrac{\pi}{6}, \tfrac{2\pi}{3}) \qquad \triangle(\tfrac{\pi}{3}, \tfrac{\pi}{3}, \tfrac{\pi}{3})$$

Fig. 1. A generating chain

In this paper, we consider mainly the 2 dimensional case. Our results strongly depend on the works of Laczkovich [7,8]. We show the following.

- $e(\{3\}, \{n\}) = 2$ for every $n > 3$, $n \neq 6$.
  $e(\{4\}, \{n\}) = 2$ for every $n > 4$.
  $e(\{3\}, \{4\}, \{6\}, \{12\}) = 2$.
  If $m, n\, (m \neq n)$ are sufficiently large, then $e(\{m\}, \{n\}) = 2$.
- All right triangles are original.
  Isosceles triangles other than $\triangle(\tfrac{\pi}{6}, \tfrac{\pi}{6}, \tfrac{2\pi}{3})$ are all terminal.
  Intermediate triangles are only the following types:

$$\triangle(\tfrac{\pi}{6}, \tfrac{\pi}{6}, \tfrac{2\pi}{3}),\ \triangle(\alpha, 2\alpha, \pi - 3\alpha), \triangle(\beta, \tfrac{\pi}{3}, \tfrac{2\pi - 3\beta}{3}),$$

  where α, β take suitable irrational multiples of π (infinitely many values are possible for α, β).
- The generating chain $(*)$ is the only chain of length ≥ 3 that contains the intermediate triangle $\triangle(\tfrac{\pi}{6}, \tfrac{\pi}{6}, \tfrac{2\pi}{3})$.
  The length of every generating chain of triangles is at most 3.

2   Polygons

Theorem 2.1. *If a polygon P has n sides whose lengths are linearly independent over the rational field $\mathbb{Q}$, then $e(P) \geq n/3$.*

Proof. Let $a_1, a_2, \ldots, a_n$ be the side-lengths of P that are linearly independent over $\mathbb{Q}$. Suppose that a set of k triangles tiles P, and let $b_1, b_2, \ldots, b_{3k}$ be the side-lengths of the k triangles. Each a_i can be represented as a linear combination of $b_1, \ldots, b_{3k}$ with integral coefficients. Since $a_1, \ldots, a_n$ are linearly independent over $\mathbb{Q}$, we must have $3k \geq n$, and hence $k \geq n/3$.   □

Corollary 2.1. *For any integer $n > 0$, there is a polygon P with $e(P) \geq n$.*

Example 2.1. *The simplicial element number of a quadrilateral with sides $1, \sqrt{2}, \sqrt{3}, \sqrt{5}$ is 2.*

The following theorem is proved by Laczkovich [8, Theorems 3.3, 3.6].

Theorem 2.2 (Laczkovich 2012).

(i) *A triangle T can generate a rectangle $a \times b$ if and only if T is a right triangle in which the ratio of perpendicular sides is a rational multiple of a/b or b/a.*

(ii) *A triangle T can generate an equilateral triangle if and only if one of the following holds.*
   – *T is $\triangle(\frac{\pi}{6}, \frac{\pi}{3}, \frac{\pi}{2})$.*
   – *T is $\triangle(\frac{\pi}{6}, \frac{\pi}{6}, \frac{2\pi}{3})$.*
   – *The ratios of sides of T are all rationals, and one angle of T is $\pi/3$ or $2\pi/3$.*

Example 2.2. If the side-lengths of T are $7, 8, 13$ (in this case, the largest angle of T becomes $2\pi/3$), then the equilateral triangle of side 11760 can be dissected into 2469600 triangles each congruent to T, see Laczkovich [7] p.86.

Let us state a well-known fact (see Appendix D of Niven [10]) as a lemma for later use.

Lemma 2.1. *Let α be an acute angle of a triangle. If $\cos\alpha \in \mathbb{Q}$, then either α/π is irrational or $\alpha = \pi/3$. If $\tan\alpha \in \mathbb{Q}$, then either α/π is irrational or $\alpha = \pi/4$.*   □

From Theorem 2.2, we have the following.

Theorem 2.3. (i) *For any $n > 3, n \neq 6$, an equilateral triangle and a regular n-gon are independent.* (ii) *A square and any other regular polygon are independent.*

Proof. (i) First, note that the interior angle of a regular n-gon is $(n-2)\pi/n$. Let T be a triangle that generates an equilateral triangle. We use Theorem 2.2 (ii). If T is $\triangle(\frac{\pi}{6}, \frac{\pi}{6}, \frac{2\pi}{3})$ or $\triangle(\frac{\pi}{6}, \frac{\pi}{3}, \frac{\pi}{2})$, then, except the cases $n = 4, 6, 12$, no angle $(n-2)\pi/n$ can be constructed as a linear combination of the angles of T with nonnegative integral coefficients. By Theorem 2.2 (i), T cannot generate $\{4\}$. Suppose that T generates $\{12\}$. Since $\triangle(\frac{\pi}{6}, \frac{\pi}{3}, \frac{\pi}{2})$ generates $\triangle(\frac{\pi}{6}, \frac{\pi}{6}, \frac{2\pi}{3})$, we may suppose that T is a triangle with sides $1, \sqrt{3}, 2$ (which is a $\triangle(\frac{\pi}{6}, \frac{\pi}{3}, \frac{\pi}{2})$) and T tiles a regular 12-gon P. The side-length s of P can be then represented as $s = a + b\sqrt{3}$ (where a, b are nonnegative integers), and the area of P is

$$12 \times (s/2)^2 \cot \tfrac{\pi}{12} = 3s^2(2 + \sqrt{3}) = 3(a + b\sqrt{3})^2(2 + \sqrt{3})$$
$$= 3(2a^2 + 6b^2 + 6ab) + 3(4ab + a^2 + 3b^2)\sqrt{3}.$$

On the other hand, since T tiles P, the area of P is an integral multiple of $\sqrt{3}/2$, which cannot be equal to $3(2a^2+6b^2+6ab)+3(4ab+a^2+3b^2)\sqrt{3}$, a contradiction.

Now, suppose that the ratios of the sides of T are all rationals and T has an angle equal to $\pi/3$ or $2\pi/3$. If α denotes the smallest angle of T, then $\cos\alpha$ is a rational by the law of cosines. Hence α/π is irrational or $\alpha = \pi/3$ by Lemma 2.1. If $\alpha = \pi/3$, then T is an equilateral triangle, and no angle $(n-2)\pi/n$ $(n \neq 6)$ is an integral multiple of $\pi/3$. If $\alpha < \pi/3$, then α/π is irrational, and by Theorem 2.2 (ii), T must have $2\pi/3$ as its largest angle. In this case, no angle $(n-2)\pi/n$ $(n \neq 6)$ can be constructed as a linear combination of the angles of T with nonnegative integral coefficients.

(ii) Let T be a triangle that generates a square, and let α be the smallest angle of T. By Theorem 2.2 (i), T is a right triangle and $\tan\alpha$ is a rational. Hence α/π is irrational or $\alpha = \pi/4$ by Lemma 2.1. If T generates a regular n-gon for $n \neq 3, 4$, then the angle $(n-2)\pi/n$ must be represented as a linear combination of the angles of T with nonnegative integral coefficients, which is possible only when $\alpha = \pi/4$ and $n = 8$. So, assume that $\triangle(\tfrac{\pi}{4}, \tfrac{\pi}{4}, \tfrac{\pi}{2})$ with sides $1, 1, \sqrt{2}$ can tile a regular octagon $\{8\}$, and let s denote the side-length of the octagon. This s must be written as $s = a + b\sqrt{2}$ with some nonnegative integers a, b. Let r be the circum-radius of the octagon. By the cosine law, we have $s^2 = 2r^2 - 2r^2 \cos\tfrac{\pi}{4} = r^2(2 - \sqrt{2})$, and hence $r^2 = s^2/(2 - \sqrt{2})$. The area of the octagon is $4r^2 \sin\tfrac{\pi}{4} = 4s^2/(2\sqrt{2} - 2) = 2s^2/(\sqrt{2} - 1) = 2(a + b\sqrt{2})^2/(\sqrt{2} - 1)$. Since a, b are nonnegative integers, this is an irrational number. On the other hand, since we assumed that the right isosceles triangle tiles the octagon, the area of the octagon must be an integral multiple of $1/2$, a rational number, which is a contradiction. $\qquad\square$

Example 2.3. $e(\{3\}, \{4\}, \{6\}, \{12\}) = 2$.

This can be seen as follows. Since $\{3\}, \{4\}$ are independent, the simplicial element number is at least 2. Let T_1 be the right triangle with sides $1, 2, \sqrt{3}$, and T_2 be the right triangle with perpendicular sides $1, 2 - \sqrt{3}$. Now, T_1 can generate $\{3\}$ and $\{6\}$; T_2 can generate $\{12\}$, see Fig. 2. Since $\{T_1, T_2\}$ can generate a rectangle 1×2, $\{T_1, T_2\}$ can generate a square. Hence the simplicial element number is at most 2.

If m, n are sufficiently large and $m \neq n$, then $\{m\}, \{n\}$ are independent.

Theorem 2.4. *If* $420 < m < n$, *then* $e(\{m\}, \{n\}) = 2$.

Proof. It is obvious that $e(\{m\}, \{n\}) \leq 2$. By [8, Theorem 3.4 (iii)], if $k > 420$ and $T \to \{k\}$, then T is either $\triangle(\tfrac{(k-2)\pi}{2k}, \tfrac{(k-2)\pi}{2k}, \tfrac{2\pi}{k})$ or $\triangle(\tfrac{(k-2)\pi}{2k}, \tfrac{\pi}{k}, \tfrac{\pi}{2})$. From this we can deduce that there is no triangle T such that $T \to \{m\}$ and $T \to \{n\}$. Therefore $e(\{m\}, \{n\}) \geq 2$. $\qquad\square$

Conjecture 2.1. $e(\{m\}, \{n\}) = 2$ for $3 \leq m < n$ and $(m, n) \neq (3, 6)$.

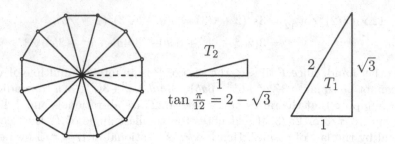

Fig. 2. Triangles T_1 and T_2

3 Triangles

Since the minimum angle in a triangle is at most $\pi/3$, we have the following.

Example 3.1. *An equilateral triangle is terminal.*

The following result is obtained by Laczkovich [7, Theorem 5.3].

Theorem 3.1 (Laczkovich 1995). *Suppose that all angles of T are rational multiples of π. (i) If the three angles of T are all different, then T is original. (ii) If T is an isosceles (not an equilateral) triangle with base angle θ, then $\sigma \to T$ implies either $\sigma = \triangle(\theta, \frac{\pi}{2} - \theta, \frac{\pi}{2})$ or $\sigma \sim T$.*

As a corollary, we have the following.

Corollary 3.1. *Suppose that all angles of T are rational multiples of π and all angles are different. If T is not a right triangle, then T is terminal.*

Proof. If $T \to \tau$, then the angles of τ are all rational multiples of π. If all angles of τ are different, then τ is original by Theorem 3.1 (i), and hence $T \sim \tau$. If just two angles of τ are equal, then since T is not a right triangle, we have $T \sim \tau$ by Theorem 3.1 (ii). This is impossible. Now, suppose that τ is an equilateral triangle. By Theorem 2.2 (ii), the ratios of the sides of T are all rational numbers. Let α be the minimum angle of T. Since T is not an equilateral triangle, it follows that $\alpha < \pi/3$. Since the ratios of side-lengths of T are all rational numbers, $\cos \alpha \in \mathbb{Q}$ by the cosine law. This implies either α/π is an irrational or $\alpha = \pi/3$ by Lemma 2.1, which is a contradiction. ◻

The following was essentially proved in Soifer [11].

Theorem 3.2. *If the three interior angles α, β, γ of a triangle T are linearly independent over $\mathbb{Q}$, then T is original and terminal.*

Proof. Suppose that T can be generated by a triangle τ with three interior angles x, y, z. Then, for some nonnegative integers l_i, m_i, n_i,

$$\begin{cases} \alpha &= \ell_1 x + m_1 y + n_1 z \\ \beta &= \ell_2 x + m_2 y + n_2 z \\ \gamma &= \ell_3 x + m_3 y + n_3 z. \end{cases}$$

Since α, β, γ are linearly independent over $\mathbb{Q}$, so are x, y, z. From

$$x + y + z = \alpha + \beta + \gamma = \left(\sum \ell_i\right)x + \left(\sum m_i\right)y + \left(\sum n_i\right)z$$

it follows that $1 = \sum \ell_i = \sum m_i = \sum n_i$. Changing the suffixes suitably, we may assume $\ell_1 = 1, m_2 = 1, n_3 = 1$. Then, $\alpha = x, \beta = y, \gamma = z$, and τ is similar to T. The latter part follows similarly.   □

Theorem 3.3. (i) *All right triangles are original, and* (ii) *isosceles triangles other than* $\triangle(\frac{\pi}{6}, \frac{\pi}{6}, \frac{2\pi}{3})$ *are terminal.*

Proof. (i) Let T be a right triangle and suppose $\tau \to T$. Since T can generate a rectangle, τ can generate a rectangle, and hence τ is a right triangle by Theorem 2.2. Let α, β ($\alpha \leq \beta$) be two acute angles of T, and α', β' ($\alpha' \leq \beta'$) be two acute angles of τ. We have $\alpha' \leq \alpha$ (and $\beta' \geq \beta$), for otherwise, τ cannot generate T. If $\alpha = \beta$ (i.e. $\alpha = \beta = \pi/4$), then T can generate a square, and hence τ can generate a square. In this case, $\tan \alpha'$ is a rational by Theorem 2.2 (i), and since $\pi/4$ is an integral multiple of α', α' is also a rational multiple of π. This implies $\alpha' = \pi/4$ by Lemma 2.1, and hence $\tau \sim T$. So, we may suppose $\alpha < \beta$. If α is a rational multiple of π, then so is β, and T is original by Theorem 3.1 (i), and $\tau \sim T$. So, we suppose α/π is not a rational. In this case, α and $\beta (= \pi/2 - \alpha)$ are linearly independent over $\mathbb{Q}$. If $\alpha' < \alpha$, then $\beta' > \beta$, and hence both α, β must be multiples of α', but this is impossible since α, β are linearly independent over $\mathbb{Q}$. Therefore, $\alpha' = \alpha$ and $\tau \sim T$.

(ii) Since an equilateral triangle is terminal, we consider the case that T is an isosceles triangle that is neither equilateral nor $\triangle(\frac{\pi}{6}, \frac{\pi}{6}, \frac{2\pi}{3})$. Let α, α, β be the three angles of T. If α or β is a rational multiple of π, then both α, β are rational multiples of π, and T is terminal by Theorem 3.1 (ii). So, suppose that one of α, β are irrational multiples of π. Then, since $2\alpha + \beta = \pi$, it follows that α, β are both irrational multiple of π, and α, β are linearly independent over $\mathbb{Q}$. Suppose that $T \to \tau$, and let x, y, z be the angles of τ. We have

$$\begin{cases} x = m_1\alpha + n_1\beta \\ y = m_2\alpha + n_2\beta \\ z = m_3\alpha + n_3\beta. \end{cases}$$

where m_i, n_i are all nonnegative integers. Hence

$$2\alpha + \beta = \pi = x + y + z = \left(\sum m_i\right)\alpha + \left(\sum n_i\right)\beta,$$

and $\sum m_i = 2, \sum n_i = 1$. Hence, after suitable change of notations, we can deduce that $x = \alpha, y = \alpha, z = \beta$. Therefore $T \sim \tau$, and T is terminal.   □

Angles of a triangle are *commensurable* if they are all rational multiples of π.

Corollary 3.2. (i) *Among the triangles with commensurable angles, the triangle* $\triangle(\frac{\pi}{6}, \frac{\pi}{6}, \frac{2\pi}{3})$ *is a unique intermediate triangle.* (ii) *The generating chain* (∗) *is a unique chain of length* ≥ 3 *that contains* $\triangle(\frac{\pi}{6}, \frac{\pi}{6}, \frac{2\pi}{3})$.

Proof. (i) By Corollary 3.1 and Theorem 3.3, it follows that among the triangles with commensurable angles, $\triangle(\frac{\pi}{6}, \frac{\pi}{6}, \frac{2\pi}{3})$ is only one intermediate triangle. (ii) Let $\sigma = \triangle(\frac{\pi}{6}, \frac{\pi}{6}, \frac{2\pi}{3})$ and suppose that $\sigma \to \tau, \sigma \not\sim \tau$. Then the angles of τ are also commensurable and τ is not a right triangle by Theorem 3.3(i). Since σ is an intermediate triangle, and $\sigma \not\sim \tau$, it follows from Theorem 3.1(i)(ii) that τ is an equilateral triangle, which is terminal. Suppose now $\eta \to \sigma, \eta \not\sim \sigma$. Then by Theorem 3.1(ii), we have $\eta = \triangle(\frac{\pi}{6}, \frac{\pi}{3}, \frac{\pi}{2})$, which is original by Theorem 3.3. Hence (∗) is a unique generating chain of length ≥ 3 that contains $\triangle(\frac{\pi}{6}, \frac{\pi}{6}, \frac{2\pi}{3})$. □

Let σ be a triangle with non-commensurable angles, and suppose that $\sigma \to \tau$ and $\sigma \not\sim \tau$. From Laczkovich [7, Theorems 4.1], all possible candidates for such pair (σ, τ) are derived as in Table 1.

The next theorem follows from Table 1.

Table 1. Possible pairs (σ, τ) when angles of σ are non-commensurable

	σ	τ
1.	$\triangle(\alpha, \frac{\pi-2\alpha}{2}, \frac{\pi}{2})$	$\triangle(\alpha, \alpha, \pi - 2\alpha)$
2.	$\triangle(\alpha, \frac{\pi}{3}, \frac{2\pi-3\alpha}{3})$	$\triangle(\frac{\pi}{3}, \frac{\pi}{3}, \frac{\pi}{3})$
3.	$\triangle(\alpha, 2\alpha, \pi - 3\alpha)$	$\triangle(\alpha, \alpha, \pi - 2\alpha)$
4.	$\triangle(\alpha, \frac{\pi-3\alpha}{2}, \frac{\pi+\alpha}{2})$	$\triangle(\alpha, \alpha, \pi - 2\alpha)$ or
		$\triangle(3\alpha, \frac{\pi-3\alpha}{2}, \frac{\pi-3\alpha}{2})$ or
		$\triangle(\alpha, 2\alpha, \pi - 3\alpha)$ or
		$\triangle(\alpha, \frac{\pi-\alpha}{2}, \frac{\pi-\alpha}{2})$ or
		$\triangle(2\alpha, \frac{\pi-\alpha}{2}, \frac{\pi-3\alpha}{2})$
5.	$\triangle(\alpha, \frac{\pi-3\alpha}{3}, \frac{2\pi}{3})$	$\triangle(\alpha, \alpha, \pi - 2\alpha)$ or
		$\triangle(\alpha, 2\alpha, \pi - 3\alpha)$ or
		$\triangle(\alpha, \frac{\pi+3\alpha}{3}, \frac{2\pi-6\alpha}{3})$ or
		$\triangle(\alpha, \frac{\pi}{3}, \frac{2\pi-3\alpha}{3})$ or
		$\triangle(2\alpha, \frac{\pi}{3}, \frac{2\pi-6\alpha}{3})$ or
		$\triangle(\frac{\pi}{3}, \frac{\pi}{3}, \frac{\pi}{3})$

Theorem 3.4. *If σ is a triangle with non-commensurable angles, then all possible types of generating chains $\eta \to \sigma \to \tau$ of length three are given as follows:*

(a) $\triangle(\alpha, \frac{\pi-3\alpha}{2}, \frac{\pi+\alpha}{2}) \to \triangle(\alpha, 2\alpha, \pi - 3\alpha) \to \triangle(\alpha, \alpha, \pi - 2\alpha)$

(b) $\triangle(\alpha, \frac{\pi-3\alpha}{3}, \frac{2\pi}{3}) \to \triangle(\alpha, 2\alpha, \pi - 3\alpha) \to \triangle(\alpha, \alpha, \pi - 2\alpha)$

(c) $\triangle(\alpha, \frac{\pi-3\alpha}{3}, \frac{2\pi}{3}) \to \triangle(\alpha, \frac{\pi}{3}, \frac{2\pi-3\alpha}{3}) \to \triangle(\frac{\pi}{3}, \frac{\pi}{3}, \frac{\pi}{3})$

(d) $\triangle(\alpha, \frac{\pi-3\alpha}{3}, \frac{2\pi}{3}) \to \triangle(2\alpha, \frac{\pi}{3}, \frac{2\pi-6\alpha}{3}) \to \triangle(\frac{\pi}{3}, \frac{\pi}{3}, \frac{\pi}{3})$.

Remark 3.1. If $\alpha < \frac{\pi}{3}$ and $\sin\frac{\alpha}{2} \in \mathbb{Q}$, then (a) is indeed a generating chain, see Laczkovich [7, Theorem 2.4]. Notice that there are infinitely many such α. If $\alpha < \frac{\pi}{3}$ and $\sqrt{3}\sin\alpha, \cos\alpha \in \mathbb{Q}$, then $(b), (c), (d)$ are all generating chains, see Laczkovich [7, Theorem 2.5], and Theorem 2.2(ii). If we let α be the smallest angle of the right triangle with sides $n^2 + 3, n^2 - 3, 2\sqrt{3}n$ $(n > 4)$, then $\alpha < \frac{\pi}{3}$ and $\sqrt{3}\sin\alpha, \cos\alpha \in \mathbb{Q}$. Hence there are also infinitely many α for generating chains $(b), (c), (d)$.

Theorem 3.5. *Intermediate triangles other than* $\triangle(\frac{\pi}{6}, \frac{\pi}{6}, \frac{2\pi}{3})$ *are only of the following types:*

$$\triangle(\alpha, 2\alpha, \pi - 3\alpha), \triangle(\beta, \frac{\pi}{3}, \frac{2\pi}{3} - \beta),$$

where α, β *take suitable irrational multiples of* π, *and infinitely many different values are possible for* α, β.

Proof. If an intermediate triangle has commensurable angles, then, by Corollary 3.2, it is $\triangle(\frac{\pi}{6}, \frac{\pi}{6}, \frac{2\pi}{3})$. If it has non-commensurable angles, then, by Theorem 3.4, it must be one of the types $\triangle(\alpha, 2\alpha, \pi - 3\alpha), \triangle(\beta, \frac{\pi}{3}, \frac{2\pi}{3} - \beta)$, and infinitely many values are possible for α, β by Remark 3.1. □

Problem 3.1. *Determine all possible values of* α *and* β *of the above theorem.*

Problem 3.2. *Determine the original triangles, terminal triangles, and intermediate triangles completely.*

Theorem 3.6. *The length of a generating chain of triangles is at most 3.*

Proof. Suppose that there exists a generating chain $\sigma_1 \to \sigma_2 \to \sigma_3 \to \sigma_4$ of length 4. Then the sub-chain $\sigma_1 \to \sigma_2 \to \sigma_3$ is either $(*)$ or a type of $(a), (b), (c), (d)$ of Theorem 3.4. Hence σ_3 is an isosceles triangle other than $\triangle(\frac{\pi}{6}, \frac{\pi}{6}, \frac{2\pi}{3})$, and hence σ_3 is terminal by Theorem 3.3 (ii). Therefore σ_4 cannot exist, a contradiction. □

4 In Higher Dimensions

Let us state here some known results in dimension ≥ 3. A d-dimensional *Hill-simplex* [6] (or Hadwiger-Hill simplex [5]) of angle θ, denoted by $Q^d(\theta)$, is defined as the convex hull of the vectors $0, v_1, v_1 + v_2, \ldots, v_1 + v_2 + \cdots + v_d$, where $v_1, \ldots, v_d$ are linearly independent unit vectors such that the angle between every two of them is equal to θ. A Hill-simplex of angle $\pi/2$ is called an *orthogonal* Hill-simplex, and a 3-dimensional Hill-simplex is called a Hill-tetrahedron. Figure 3 shows the orthogonal Hill-tetrahedron $Q^3(\pi/2)$.

For every $d \geq 3$, by bisecting $Q^d(\pi/2)$ successively, we can get a generating chain of d-simplices

$$\sigma_0 \to \sigma_1 \to \sigma_2 \to \cdots \to \sigma_d$$

such that $\sigma_d = Q^d(\pi/2)$ and σ_0 is similar to σ_d, see for the details, Maehara [9]. Since $\sigma_0 \sim \sigma_d$, we can extend this generating chain to both direction, and we can get arbitrarily long generating chain. (Cf. Theorem 3.6.)

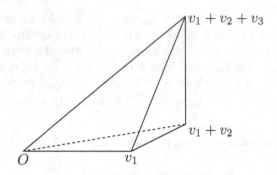

Fig. 3. The orthogonal Hill-tetrahedron $Q^3(\pi/2)$

A 3-dimensional convex polytope is called a *parallelohedron* if it tiles $\mathbb{R}^3$ by translations only. Akiyama et al. [1] proved that for any parallelohedron Π, there is an affine transformation f of $\mathbb{R}^3$ such that $f(\Pi)$ is generated by the orthogonal Hill-tetrahedron $Q^3(\pi/2)$.

The simplicial element numbers of the families of regular d-polytopes for $d \geq 2$ are determined by Akiyama et al. [2–4] as shown in Table 2. (Their "element number" is slightly different from our simplicial element number, but Table 2 follows from their works.)

Table 2. e(the d-dimensional regular polytopes)

d	# of regular polytopes	simplicial element number
2	∞	∞
3	5	4
4	6	4
≥ 5	3	3

Acknowledgment. Many thanks to the referees for valuable comments.

References

1. Akiyama, J., Kobayashi, M., Nakagawa, H., Nakamura, G., Sato, I.: Atoms for parallelohedra. Bolyai Society Mathematical Studies 24 (2013); Geometry-Intuitive, Discrete, and Convex, pp. 1–21
2. Akiyama, J., Hitotumatu, S., Sato, I.: Determination of the element number of the regular polytopes. Geom. Dedicata **159**, 89–97 (2012)
3. Akiyama, J., Sato, I.: The element number of the convex regular polytopes. Geom. Didicata **151**, 269–278 (2011)
4. Akiyama, J., Maehara, H., Nakamura, G., Sato, I.: Element number of the platonic solids. Geom Dedicata **145**, 181–193 (2010)

5. Hadwiger, H.: Hillsche hypertetraeder. Gazeta Math. (Lisboa) **12**, 47–48 (1951)
6. Hill, M.J.M.: Determination of the volume of certain species of tetrahedron without employing the method of limits. Proc. London Math. Soc. Ser. I **27**, 39–53 (1895)
7. Laczkovich, M.: Tilings of triangles. Discrete Math. **140**, 79–94 (1995)
8. Laczkovich, M.: Tilings of convex polygons with congruent triangles. Discrete Comput. Geom. **48**, 330–374 (2012)
9. Maehara, H.: Some replicating simplices other than Hill-simplices, Beiträge zur Algebra und Geometrie (to appear). doi:10.1007/s13366-013-0152-8
10. Niven, I.: Numbers: Rational and Irrational, MAA. Random House, New York (1961)
11. Soifer, A.: How does one cut a triangle?. Springer, New York (1990). (Center for Excellence in Mathematical Education, Colorado Springs)

Cross-Intersecting Families of Vectors

János Pach[1]($\boxtimes$) and Gábor Tardos[2]

[1] Ecole Polytechnique Fédérale de Lausanne, Lausanne, Switzerland
pach@renyi.hu
[2] Rényi Institute of Mathematics, Budapest, Hungary
tardos@renyi.hu

Abstract. Given a sequence of positive integers $p = (p_1, \ldots, p_n)$, let S_p denote the family of all sequences of positive integers $x = (x_1, \ldots, x_n)$ such that $x_i \leq p_i$ for all i. Two families of sequences (or vectors), $A, B \subseteq S_p$, are said to be r-cross-intersecting if no matter how we select $x \in A$ and $y \in B$, there are at least r distinct indices i such that $x_i = y_i$. We determine the maximum value of $|A| \cdot |B|$ over all pairs of r-cross-intersecting families and characterize the extremal pairs for $r \geq 1$, provided that $\min p_i > r + 1$. The case $\min p_i \leq r + 1$ is quite different. For this case, we have a conjecture, which we can verify under additional assumptions. Our results generalize and strengthen several previous results by Berge, Frankl, Füredi, Livingston, Moon, and Tokushige, and answers a question of Zhang. The special case $r = 1$ has also been settled recently by Borg.

1 Introduction

The Erdős-Ko-Rado theorem [EKR61] states that for $n \geq 2k$, every family of pairwise intersecting k-element subsets of an n-element set consists of at most $\binom{n-1}{k-1}$ subsets, as many as the star-like family of all subsets containing a fixed element of the underlying set. This was the starting point of a whole new area within combinatorics: extremal set theory; see [GK78, Bol86, DeF83, F95]. The Erdős-Ko-Rado theorem has been extended and generalized to other structures: to multisets, divisors of an integer, subspaces of a vector space, families of permutations, etc. It was also generalized to "cross-intersecting" families, i.e., to families A and B with the property that every element of A intersects all elements of B; see Hilton [Hi77], Moon [Mo82], and Pyber [Py86].

For any positive integer k, we write $[k]$ for the set $\{1, \ldots, k\}$. Given a sequence of positive integers $p = (p_1, \ldots, p_n)$, let

$$S_p = [p_1] \times \cdots \times [p_n] = \{(x_1, \ldots, x_n) \ : \ x_i \in [p_i] \text{ for } i \in [n]\}.$$

J. Pach is supported by OTKA under ERC projects GraDR and ComPoSe 10-EuroGIGA-OP-003, and by Swiss National Science Foundation Grants 200020-144531 and 200021-137574. G. Tardos is supported by OTKA grant NN-102029, an NSERC Discovery grant, the "Lendület" project of the Hungarian Academy of Sciences and by EPFL.

© Springer International Publishing Switzerland 2014
J. Akiyama et al. (Eds.): JCDCGG 2013, LNCS 8845, pp. 122–137, 2014.
DOI: 10.1007/978-3-319-13287-7_11

We will refer to the elements of S_p as *vectors*. The *Hamming distance* between the vectors $x, y \in S_p$ is $|\{i \in [n] : x_i \neq y_i\}|$ and is denoted by $d(x, y)$. Let $r \geq 1$ be an integer. Two vectors $x, y \in S_p$ are said to be r-*intersecting* if $d(x, y) \leq n - r$. (This term originates in the observation that if we represent a vector $x = (x_1, \ldots, x_n) \in S_p$ by the set $\{(i, x_i) : i \in [n]\}$, then x and $y \in S_p$ are r-intersecting if and only if the sets representing them have at least r common elements.) Two families $A, B \subseteq S_p$ are r-*cross-intersecting*, if every pair $x \in A$, $y \in B$ is r-intersecting. If (A, A) is an r-cross-intersecting pair, we say A is r-*intersecting*. We simply say *intersecting* or *cross-intersecting* to mean 1-intersecting or 1-cross-intersecting, respectively.

The investigation of the maximum value for $|A| \cdot |B|$ for cross-intersecting pairs of families $A, B \subseteq S_p$ was initiated by Moon [Mo82]. She proved, using a clever induction argument, that in the special case when $p_1 = p_2 = \cdots = p_n = k$ for some $k \geq 3$, every cross-intersecting pair $A, B \subseteq S_p$ satisfies

$$|A| \cdot |B| \leq k^{2n-2},$$

with equality if and only if $A = B = \{x \in S_p : x_i = j\}$, for some $i \in [n]$ and $j \in [k]$. In the case $A = B$, Moon's theorem had been discovered by Berge [Be74], Livingston [Liv79], and Borg [Bo08]. See also Stanton [St80]. In his report on Livingston's paper, published in the *Mathematical Reviews*, Kleitman gave an extremely short proof for the case $A = B$, based on a shifting argument. Zhang [Zh13] established a somewhat weaker result, using a generalization of Katona's circle method [K72]. Note that for $k = 2$, we can take $A = B$ to be any family of 2^{n-1} vectors without containing a pair $(x_1, \ldots, x_n), (y_1, \ldots, y_n)$ with $x_i + y_i = 3$ for every i. Then A is an intersecting family with $|A|^2 = 2^{2n-2}$, which is not of the type described in Moon's theorem.

Moon also considered r-cross-intersecting pairs in S_p with $p_1 = p_2 = \cdots = p_n = k$ for some $k > r + 1$, and characterized all pairs for which $|A| \cdot |B|$ attains its maximum, that is, we have

$$|A| \cdot |B| = k^{2(n-r)}.$$

The assumption $k > r + 1$ is necessary. See Tokushige [To13], for a somewhat weaker result, using algebraic techniques.

Zhang [Zh13] suggested that Moon's results may be extended to arbitrary sequences of positive integers $p = (p_1, \ldots, p_n)$. The aim of this note is twofold: (1) to establish such an extension under the assumption $\min_i p_i > r + 1$, and (2) to formulate a conjecture that covers essentially all other interesting cases. We verify this conjecture in several special cases.

We start with the case $r = 1$, which has also been settled by Borg [Bo14], using different techniques.

Theorem 1. *Let $p = (p_1, \ldots, p_n)$ be a sequence positive integers and let $A, B \subseteq S_p$ form a pair of cross-intersecting families of vectors.*

We have $|A| \cdot |B| \leq |S_p|^2/k^2$, where $k = \min_i p_i$. Equality holds for the case $A = B = \{x \in S_p : x_i = j\}$, whenever $i \in [n]$ satisfies $p_i = k$ and $j \in [k]$. For $k \neq 2$, there are no other extremal cross-intersecting families.

We say that a coordinate $i \in [n]$ is *irrelevant* for a set $A \subseteq S_p$ if, whenever two elements of S_p differ only in coordinate i and A contains one of them, it also contains the other. Otherwise, we say that i is *relevant* for A.

Note that no coordinate i with $p_i = 1$ can be relevant for any family. Each such coordinate forces an intersection between every pair of vectors. So, if we delete it, every r-cross-intersecting pair becomes $(r-1)$-cross-intersecting. Therefore, from now on we will always assume that we have $p_i \geq 2$ for every i.

We call a sequence of integers $p = (p_1, \ldots, p_n)$ a *size vector* if $p_i \geq 2$ for all i. The *length* of p is n. We say that an r-cross-intersecting pair $A, B \subseteq S_p$ is *maximal* if it maximizes the value $|A| \cdot |B|$.

Using this notation and terminology, Theorem 1 can be rephrased as follows.

Theorem 1'. *Let $p = (p_1, \ldots, p_n)$ be a sequence of positive integers with $k = \min_i p_i > 2$.*

For any maximal pair of cross-intersecting families, $A, B \subseteq S_p$, we have $A = B$, and there is a single coordinate which is relevant for A. The relevant coordinate i must satisfy $p_i = k$.

See Sect. 5 for a complete characterization of maximal cross-intersecting pairs in the $k = 2$ case. Here we mention that only the coordinates with $p_i = 2$ can be relevant for them, but for certain pairs, *all* such coordinates are relevant simultaneously. For example, let n be odd, $p = (2, \ldots, 2)$, and let $A = B$ consist of all vectors in S_p which have at most $\lfloor n/2 \rfloor$ coordinates that are 1. This makes (A, B) a maximal cross-intersecting pair.

Let $T \subseteq [n]$ be a subset of the coordinates, let $x_0 \in S_p$ be an arbitrary vector, and let k be an integer satisfying $0 \leq k \leq |T|$. The *Hamming ball* of radius k around x_0 in the coordinates T is defined as the family

$$B_k = \{x \in S_p \; : \; |\{i \in T \; : \; x_i \neq (x_0)_i\}| \leq k\}.$$

Note that the pair (B_k, B_l) is $(|T| - k - l)$-cross-intersecting. We use the word *ball* to refer to any Hamming ball without specifying its center, radius or its set of coordinates. A Hamming ball of radius 0 in coordinates T is said to be obtained by *fixing* the coordinates in T.

For the proof of Theorem 1, we need the following statement, which will be established by induction on n, using the idea in [Mo82].

Lemma 2. *Let $1 \leq r < n$, let $p = (p_1, \ldots, p_n)$ be a size vector satisfying $3 \leq p_1 \leq p_2 \leq \cdots \leq p_n$ and let $A, B \subseteq S_p$ form a pair of r-cross-intersecting families. If*

$$\frac{2}{p_{r+1}} + \sum_{i=1}^{r} \frac{1}{p_i} \leq 1,$$

then $|A| \cdot |B| \leq \prod_{i=r+1}^{n} p_i^2$. In case of equality, we have $A = B$ and this family can be obtained by fixing r coordinates in S_p.

By fixing any r coordinates, we obtain a "trivial" r-intersecting family $A = B \subseteq S_p$. As was observed by Frankl and Füredi [FF80], not all maximal size

r-intersecting families can be obtained in this way, for certain size vectors. They considered size vectors $p = (k, \ldots, k)$ with $n \geq r + 2$ coordinates, and noticed that a Hamming ball of radius 1 in $r + 2$ coordinates is r-intersecting. Moreover, for $k \leq r$, this family is strictly larger than the trivial r-intersecting family. See also [AhK98].

On the other hand, as was mentioned before, for $k \geq r + 2$, Moon [Mo82] proved that among all r-intersecting families, the trivial ones are maximal.

This leaves open only the case $k = r + 1$, where the trivial r-intersecting families and the radius 1 balls in $r + 2$ coordinates have precisely the same size. We believe that in this case there are no larger r-intersecting families. For $r = 1$, it can be and has been easily verified (and follows, for example, from our Theorem 1, which deals with the asymmetric case, when A and B do not necessarily coincide). Our Theorem 7 settles the problem also for $r > 6$. The intermediate cases $2 \leq r \leq 6$ are still open, but they could possibly be handled by computer search.

Therefore, to characterize maximal size r-intersecting families A or maximal r-cross-intersecting pairs of families (A, B) for all size vectors, we cannot restrict ourselves to fixing r coordinates. We make the following conjecture that can roughly be considered as a generalization of the Frankl-Füredi conjecture [FF80] that has been proved by Frankl and Tokushige [FT99]. The generalization is twofold: we consider r-cross-intersecting pairs rather than r-intersecting families and we allow arbitrary size vectors not just vectors with all-equal coordinates.

Conjecture 3. *If $1 \leq r \leq n$ and p is a size vector of length n, then there exists a maximal pair of r-cross-intersecting families $A, B \subseteq S_p$, where A and B are balls. If we further have $p_i \geq 3$ for all $i \in [n]$, then all maximal pairs of r-cross-intersecting families consist of balls.*

Note that the $r = 1$ special case of Conjecture 3 is established by Theorem 1. Some further special cases of the conjecture are settled in Theorem 7.

It is not hard to narrow down the range of possibilities for maximal r-cross-intersecting pairs that are formed by two balls, A and B. In fact, the following simple lemma implies that all such pairs are determined up to isomorphism by the radii of A and B. Assuming that Conjecture 3 is true, finding $\max |A| \cdot |B|$ for r-cross-intersecting families $A, B \subseteq S_p$ boils down to making numeric comparisons for pairs of balls obtained by all possible radii. In case $p_i \geq 3$ for all i, the same process also finds all maximal r-cross-intersecting pairs.

Lemma 4. *Let $1 \leq r \leq n$ and let $p = (p_1, \ldots, p_n)$ be a size vector. If $A, B \subseteq S_p$ form a maximal pair of r-cross-intersecting families, then either of them determines the other. In particular, A and B have the same set of relevant coordinates. Moreover, if A is a ball of radius l around $x_0 \in S_p$ in a set of coordinates $T \subseteq [n]$, then $|T| \geq l + r$, B is a ball of radius $|T| - l - r$ around x_0 in the same set of the coordinates, and we have $p_i \leq p_j$ for $i \in T$ and $j \in [n] \setminus T$.*

As we have indicated above, we have been unable to prove Conjecture 3 in its full generality, but we were able to verify it in several interesting special cases.

We will proceed in two steps. First we argue, using *entropies*, that the number of relevant coordinates in a maximal r-cross-intersecting family is bounded. Then we apply combinatorial methods to prove the conjecture under the assumption that the number of relevant coordinates is small.

In the case when there are many relevant coordinates for a pair of maximal r-cross-intersecting families, we use entropies to bound the size of the families and to prove

Theorem 5. *Let $1 \le r \le n$, let $p = (p_1, \ldots, p_n)$ be a size vector, let $A, B \subseteq S_p$ form a maximal pair of r-cross-intersecting families and let T be the set of coordinates that are relevant for A or B. Then neither the size of A nor the size of B can exceed*

$$\frac{|S_p|}{\prod_{i \in T}(p_i - 1)^{1 - 2/p_i}}.$$

We use this theorem to bound the number of relevant coordinates i with $p_i > 2$. The number of relevant coordinates i with $p_i = 2$ can be unbounded; see Sect. 5.

Theorem 6. *Let $1 \le r \le n$, let $p = (p_1, \ldots, p_n)$ be a size vector and let $A, B \subseteq S_p$ form a maximal pair of r-cross-intersecting families.*

For the set of coordinates T relevant for A or B, we have

$$\prod_{i=1}^{r} p_i \ge \prod_{i \in T}(p_i - 1)^{1 - 2/p_i},$$

which implies that $|\{i \in T : p_i > 2\}| < 5r$.

We can characterize the maximal r-cross-intersecting pairs for all size vectors p satisfying $\min p_i > r + 1$, and in many other cases.

Theorem 7. *Let $1 \le r \le n$, let $p = (p_1, \ldots, p_n)$ be a size vector with $p_1 \le p_2 \le \cdots \le p_n$ and let $A, B \subseteq S_p$ form a pair of r-cross-intersecting families.*

1. *If $p_1 > r + 1$, we have $|A| \cdot |B| \le \prod_{i=r+1}^{n} p_i^2$. In case of equality, $A = B$ holds and this family can be obtained by fixing r coordinates in S_p.*
2. *If $p_1 = r + 1 > 7$, we have $|A| \cdot |B| \le \prod_{i=r+1}^{n} p_i^2$. In case of equality, $A = B$ holds and this family can be obtained either by fixing r coordinates in S_p or by taking a Hamming ball of radius 1 in $r + 2$ coordinates i, all satisfying $p_i = r + 1$.*
3. *There is a function $t(r) = o(r)$ such that if $p_1 \ge 2r/3 + t(r)$ and (A, B) is a maximal r-cross-intersecting pair, then the families A and B are balls of radius 0 or 1 in at most $r + 2$ coordinates.*

The proof of Theorem 7 relies on the following result.

Theorem 8. *Let $1 \le r \le n$ and let p be a size vector of length n with $p_i > 2$ for all $i \in [n]$. If $A, B \subseteq S_p$ is a maximal pair of r-cross-intersecting families and at most $r + 2$ coordinates are relevant for them, then A and B are balls of radius 0 or 1.*

With an involved case analysis, it may be possible to extend Theorem 8 to pairs with more relevant coordinates. Any such an improvement carries over to Theorem 7.

All of our results remain meaningful in the symmetric case where $A = B$. For instance, in this case, Theorem 1 (also proved by Borg [Bo14]) states that every intersecting family $A \subseteq S_p$ has at most $|S_p|/k$ members, where $k = \min_i p_i$. In case $k > 2$, equality can be achieved only by fixing some coordinate i with $p_i = k$. Note that in the case $A = B$ (i.e., r-intersecting families) the exact maximum size is known for size vectors $(q, \ldots, q)$, [FT99].

2   Proof of Theorem 1

The aim of this section is to establish Theorem 1. First, we verify Lemma 4 and another technical lemma (see Lemma 9 below), which generalizes the corresponding result in [Mo82]. Our proof is slightly simpler. Lemma 9 will enable us to deduce Lemma 2, the main ingredient of the proof of Theorem 1, presented at the end of the section.

Proof of Lemma 4. The first statement is self-evident: if $A, B \subseteq S_p$ form a maximal pair of r-cross-intersecting families, then

$$B = \{x \in S_p \ : \ x \ r\text{-intersects } y \text{ for all } y \in A\}.$$

If a coordinate is irrelevant for A, then it is also irrelevant for B defined by this formula. Therefore, by symmetry, A and B have the same set of relevant coordinates.

If A is the Hamming ball around x_0 of radius l in coordinates T, then we have $B = \emptyset$ if $|T| < l + r$, which is not possible for a maximal cross-intersecting family. If $|T| \geq l + r$, we obtain the ball claimed in the lemma. For every $i \in T$, $j \in [n] \backslash T$, consider the set $T' = (T \backslash \{i\}) \cup \{j\}$ and the Hamming balls A' and B' of radii l and $|T| - l - r$ around x_0 in the coordinates T'. These balls form an r-cross-intersecting pair and in case $p_i > p_j$, we have $|A'| > |A|$ and $|B'| > |B|$, contradicting the maximality of the pair (A, B).   □

The following lemma will also be used in the proof of Theorem 5, presented in the next section.

Lemma 9. *Let $1 \leq r \leq n$, let $p = (p_1, \ldots, p_n)$ be a size vector, and let $A, B \subseteq S_p$ form a maximal pair of r-cross-intersecting families.*

If $i \in [n]$ is a relevant coordinate for A or B, then there exists a value $l \in [p_i]$ such that

$$|\{x \in A \ : \ x_i \neq l\}| \leq |A|/p_i,$$

$$|\{y \in B \ : \ y_i \neq l\}| \leq |B|/p_i.$$

Proof. Let us fix r, n, p, i, A, and B as in the lemma. By Lemma 4, if a coordinate is irrelevant for A, then it is also irrelevant for B and vice versa.

In the case $n = r$, we have $A = B$ and this family must be a singleton, so that the lemma is trivially true. From now on, we assume that $n > r$ and hence the notion of r-cross-intersecting families is meaningful for $n - 1$ coordinates.

Let $q = (p_1, \ldots, p_{i-1}, p_{i+1}, \ldots, p_n)$. For any $l \in [p_i]$, let

$$A_l' = \{x \in A : x_i = l\},$$

$$B_l' = \{y \in B : y_i = l\},$$

and let A_l and B_l stand for the families obtained from A_l' and B_l', respectively, by dropping their ith coordinates. By definition, we have $A_l, B_l \subseteq S_q$, and $|A| = \sum_l |A_l|$ and $|B| = \sum_l |B_l|$. Furthermore, for any two distinct elements $l, m \in [p_i]$, the families A_l and B_m are r-cross-intersecting, since the vectors in A_l' differ from the vectors in B_m' in the ith coordinate, and therefore the r indices where they agree must be elsewhere.

Let Z denote the maximum product $|A^*| \cdot |B^*|$ of an r-cross-intersecting pair $A^*, B^* \subseteq S_q$. We have $|A_l| \cdot |B_m| \leq Z$ for all $l, m \in [p_i]$ with $l \neq m$. Adding an irrelevant ith coordinate to the maximal r-cross-intersecting pair $A^*, B^* \subseteq S_q$, we obtain a pair $A^{*\prime}, B^{*\prime} \subseteq S_p$ with $|A^{*\prime}| \cdot |B^{*\prime}| = p_i^2 Z$. Thus, using the maximality of A and B, we have $|A| \cdot |B| \geq p_i^2 Z$. Let l_0 be chosen so as to maximize $|A_{l_0}| \cdot |B_{l_0}|$, and let $c = |A_{l_0}| \cdot |B_{l_0}|/Z$.

Assume first that $c \leq 1$. Then we have

$$p_i^2 Z \leq |A| \cdot |B| = \sum_{l,m \in [p_i]} |A_l| \cdot |B_m| \leq \sum_{l,m \in [p_i]} Z = p_i^2 Z.$$

Hence, we must have equality everywhere. This yields that $c = 1$ and that A_l and B_m form a maximal r-cross-intersecting pair for all $l, m \in [p_i]$, $l \neq m$. This also implies that $|A_l| = |A_m|$ for $l, m \in [p_i]$, from where the statement of the lemma follows, provided that $p_i = 2$.

If $p_i \geq 3$, then all families A_l must be equal to one another, since one member in a maximal r-cross-intersecting family determines the other (by Lemma 4). This contradicts our assumption that the ith coordinate was relevant for A.

Thus, we may assume that $c > 1$.

For $m \in [p_i]$, $m \neq l_0$, we have $|A_{l_0}| \cdot |B_m| \leq Z = |A_{l_0}| \cdot |B_{l_0}|/c$. Thus,

$$|B_m| \leq |B_{l_0}|/c, \tag{1}$$

which yields that $|B| = \sum_{m \in [p_i]} |B_m| \leq (1 + (p_i - 1)/c)|B_{l_0}|$. By symmetry, we also have

$$|A_m| \leq |A_{l_0}|/c \tag{2}$$

for $m \neq l_0$ and $|A| \leq (1 + (p_i - 1)/c)|A_{l_0}|$. Combining these inequalities, we obtain

$$p_i^2 Z \leq |A| \cdot |B| \leq (1 + (p_n - 1)/c)^2 |A_{l_0}| \cdot |B_{l_0}| = (1 + (p_i - 1)/c)^2 cZ.$$

We solve the resulting inequality $p_i^2 \leq c(1 + (p_i - 1)/c)^2$ for $c > 1$ and conclude that $c \geq (p_i - 1)^2$. This inequality, together with Eqs. (1) and (2), completes the proof of Lemma 9. $\qquad\square$

Proof of Lemma 2. We proceed by induction on n.

Let A and B form a maximal r-cross-intersecting pair. It is sufficient to show that they have only r relevant coordinates. Let us suppose that the set T of their relevant coordinates satisfies $|T| > r$, and choose a subset $T' \subseteq T$ with $|T'| = r + 1$. By Lemma 9, for every $i \in T$ there exists $l_i \in [p_i]$ such that the family $X_i = \{x \in B : x_i \neq l_i\}$ has cardinality $|X_i| \leq |B|/p_i$.

If we assume that

$$\frac{2}{p_{r+1}} + \sum_{i=1}^{r} \frac{1}{p_i} < 1$$

holds (with strict inequality), then this bound of $|X_i|$ would suffice. In order to be able to deal with the case

$$\frac{2}{p_{r+1}} + \sum_{i=1}^{r} \frac{1}{p_i} = 1,$$

we show that $|X_i| = |B|/p_i$ is not possible. Considering the proof of Lemma 9, equality here would mean that the families A_l and B_l (obtained by dropping the ith coordinate from the vectors in the sets $\{x \in A : x_i = l\}$ and $\{y \in B : y_i = l\}$, respectively) satisfy the following condition: both (A_{l_i}, B_m) and (A_m, B_{l_i}) should be maximal r-cross-intersecting pairs for all $m \neq l_i$. By the induction hypothesis, this would imply that $A_{l_i} = B_m$ and $A_m = B_{l_i}$, contradicting that $|A_m| < |A_{l_i}|$ and $|B_m| < |B_{l_i}|$ (see (1), in view of $c > 1$). Therefore, we have $|X_i| < |B|/p_i$.

Let $C = \{x \in S_p : x_i = 1 \text{ for all } i \in [r]\}$ be the r-intersecting family obtained by fixing r coordinates in S_p. In the family $D = B \backslash (\bigcup_{i \in T'} X_i)$, the coordinates in T' are fixed. Thus, we have

$$|D| \leq \prod_{i \in [n] \backslash T'} p_i \leq \prod_{i=r+2}^{n} p_i = |C|/p_{r+1}.$$

On the other hand, we have

$$|D| = |B| - \sum_{i \in T'} |X_i| > |B|(1 - \sum_{i \in T'} 1/p_i) \geq |B|(1 - \sum_{i=1}^{r+1} 1/p_i).$$

Comparing the last two inequalities, we obtain

$$|B| < \frac{|C|}{p_{r+1}(1 - \sum_{i=1}^{r+1} 1/p_i)}.$$

By our assumption on p, the denominator is at least 1, so that we have $|B| < |C|$. By symmetry, we also have $|A| < |C|$. Thus, $|A| \cdot |B| < |C|^2$ contradicting the maximality of the pair (A, B). This completes the proof of Lemma 2. $\qquad\square$

Now we can quickly finish the proof of Theorem 1.

Proof of Theorem 1. Notice that Lemma 2 implies Theorem 1, whenever $k = \min_i p_i \geq 3$. It remains to verify the statement for $k = 1$ and $k = 2$. For $k = 1$, it follows from the fact that all pairs of vectors in S_p are intersecting, thus the only maximal cross-intersecting pair is $A = B = S_p$.

Suppose next that $k = 2$. For $x \in S_p$, let $x' \in S_p$ be defined by $x'_i = (x_i + 1 \bmod p_i)$ for $i \in [n]$. Note that $x \mapsto x'$ is a permutation of S_p. Clearly, x and x' are not intersecting, so we either have $x \notin A$ or $x' \notin B$. As a consequence, we obtain that $|A| + |B| \leq |S_p|$, which, in turn, implies that $|A| \cdot |B| \leq |S_p|^2/4$, as claimed. It also follows that all maximal pairs satisfy $|A| = |B| = |S_p|/2$. □

3   Using Entropy: Proofs of Theorems 5 and 6

Proof of Theorem 5. Let r, n, p, A, B and T be as in the theorem. Let us write y for a randomly and uniformly selected element of B. Lemma 9 implies that, for each $i \in T$, there exists a value $l_i \in [p_i]$ such that

$$Pr[y_i = l_i] \geq 1 - 1/p_i. \tag{3}$$

We bound the *entropy* $H(y_i)$ of y_i from above by the entropy of the indicator variable of the event $y_i = l_i$ plus the contribution coming from the entropy of y_i assuming $y_i \neq l_i$:

$$H(y_i) \leq h(1 - 1/p_i) + (1/p_i)\log(p_i - 1) = \log p_i - (1 - 2/p_i)\log(p_i - 1),$$

where $h(z) = -z \log z - (1 - z)\log(1 - z)$ is the entropy function, and we used that $1 - 1/p_i \geq 1/2$.

For any $i \in [n]\setminus T$, we use the trivial estimate $H(y_i) \leq \log p_i$. By the subadditivity of the entropy, we obtain

$$\log|B| = H(y) \leq \sum_{i \in [n]} H(y_i) \leq \sum_{i \in T}(\log p_i - (1 - 2/p_i)\log(p_i - 1)) + \sum_{i \in [n]\setminus T} \log p_i,$$

or, equivalently,

$$|B| \leq \prod_{i \in T} \frac{p_i}{(p_i - 1)^{1 - 2/p_i}} \prod_{i \in [n]\setminus T} p_i = \frac{|S_p|}{\prod_{i \in T}(p_i - 1)^{1 - 2/p_i}}$$

as required. The bound on $|A|$ follows by symmetry and completes the proof of the theorem. □

Theorem 6 is a simple corollary of Theorem 5.

Proof of Theorem 6. Fixing the first r coordinates, we obtain the family

$$C = \{x \in S_p \: : \: x_i = 1 \text{ for all } i \in [r]\}.$$

This family is r-intersecting. Thus, by the maximality of the pair (A, B), we have

$$|A| \cdot |B| \geq |C|^2 = \left(\prod_{i=r+1}^{n} p_i \right)^2 . \tag{4}$$

Comparing this with our upper bounds on $|A|$ and $|B|$, we obtain the inequality claimed in the theorem.

To prove the required bound on the number of relevant coordinates i with $p_i \neq 2$, we assume that the coordinates are ordered, that is $p_1 \leq p_2 \leq \cdots \leq p_n$. Applying the above estimate on $\prod_{i \in [r]} p_i$ and using $(p_i - 1)^{1-2/p_i} > p_i^{1/5}$ whenever $p_i \geq 3$, the theorem follows. $\square$

4 Monotone Families: Proofs of Theorems 8 and 7

Given a vector $x \in S_p$, the set $\mathrm{supp}(x) = \{i \in [n] : x_i > 1\}$ is called the *support* of x. A family $A \subseteq S_p$ is said to be *monotone*, if for any $x \in A$ and $y \in S_p$ satisfying $\mathrm{supp}(y) \subseteq \mathrm{supp}(x)$, we have $y \in A$.

For a family $A \subseteq S_p$, let us define its *support* as $\mathrm{supp}(A) = \{\mathrm{supp}(x) : x \in A\}$. For a monotone family A, its support is clearly subset-closed and it uniquely determines A, as $A = \{x \in S_p : \mathrm{supp}(x) \in \mathrm{supp}(A)\}$.

The next result shows that if we want to prove Conjecture 3, it is sufficient to prove it for monotone families. This will enable us to establish Theorems 8 and 7, that is, to verify the conjecture for maximal r-cross-intersecting pairs with a limited number of relevant coordinates. Note that similar reduction to monotone families appears also in [FF80].

Lemma 10. *Let $1 \leq r \leq n$ and let p be a size vector of length n.*

There exists a maximal pair of r-cross-intersecting families $A, B \subseteq S_p$ such that both A and B are monotone.

If $p_i \geq 3$ for all $i \in [n]$ and $A, B \subseteq S_p$ are maximal r-cross-intersecting families that are not balls, then there exists a pair of maximal r-cross-intersecting families that consists of monotone families that are not balls and have no more relevant coordinates than A or B.

Proof. Consider the following *shift operations*. For any $i \in [n]$ and $j \in [p_i] \setminus \{1\}$, for any family $A \subseteq S_p$ and any element $x \in A$, we define

$$\phi_i(x) = (x_1, \ldots, x_{i-1}, 1, x_{i+1}, \ldots, x_n),$$

$$\phi_{i,j}(x, A) = \begin{cases} \phi_i(x) & \text{if } x_i = j \text{ and } \phi_i(x) \notin A \\ x & \text{otherwise,} \end{cases}$$

$$\phi_{i,j}(A) = \{\phi_{i,j}(x, A) : x \in A\}.$$

Clearly, we have $|\phi_{i,j}(A)| = |A|$ for any family $A \subseteq S_p$. We claim that for any pair of r-cross-intersecting families $A, B \subseteq S_p$, the families $\phi_{i,j}(A)$ and $\phi_{i,j}(B)$ are also r-cross-intersecting. Indeed, if $x \in A$ and $y \in B$ are r-intersecting vectors, then $\phi_{i,j}(x, A)$ and $\phi_{i,j}(y, B)$ are also r-intersecting, unless x and y have exactly r common coordinates, one of them is $x_i = y_i = j$, and this common coordinate gets ruined as $\phi_{i,j}(x, A) = x$ and $\phi_{i,j}(y, B) = \phi_i(y)$ (or vice versa). However, this is impossible, because this would imply that the vector $\phi_i(x)$ belongs to A, in spite of the fact that $\phi_i(x)$ and $y \in B$ are not r-intersecting.

If (A, B) is a *maximal* r-cross-intersecting pair, then so is $(\phi_{i,j}(A), \phi_{i,j}(B))$. When applying one of these shift operations does change the families A or B, then the total sum of all coordinates of all elements decreases. Therefore, after shifting a finite number of times we arrive at a maximal pair of r-intersecting families that cannot be changed by further shifting. We claim that this pair (A, B) is monotone. Let $y \in B$ and $y' \in S_p \backslash B$ be arbitrary. We show that B is monotone by showing that $\mathrm{supp}(y')$ is not contained in $\mathrm{supp}(y)$. Indeed, by the maximality of the pair (A, B) and using the fact that $y' \notin B$, there must exist $x' \in A$ such that x' and y' are not r-cross-intersecting, and hence $|\mathrm{supp}(x') \cup \mathrm{supp}(y')| > n - r$. Applying "projections" ϕ_i to x' in the coordinates $i \in \mathrm{supp}(x') \cap \mathrm{supp}(y)$, we obtain x with $\mathrm{supp}(x) = \mathrm{supp}(x') \backslash \mathrm{supp}(y)$. The shift operations $\phi_{i,j}$ do not change the family A, thus A must be closed for the projections ϕ_i and we have $x \in A$. The supports of x and y are disjoint. Thus, their Hamming distance is $|\mathrm{supp}(x) \cup \mathrm{supp}(y)|$, which is at most $n - r$, as they are r-intersecting. Therefore, $\mathrm{supp}(x) \cup \mathrm{supp}(y) = \mathrm{supp}(x') \cup \mathrm{supp}(y)$ is smaller than $\mathrm{supp}(x') \cup \mathrm{supp}(y')$, showing that $\mathrm{supp}(y') \nsubseteq \mathrm{supp}(y)$. This proves that B is monotone. By symmetry, A is also monotone, which proves the first claim of the lemma.

To prove the second claim, assume that $p_i \geq 3$ for all $i \in [n]$. Note that Theorem 1 establishes the lemma in the case $r = 1$, so from now on we can assume without loss of generality that $r \geq 2$. Let $A, B \subseteq S_p$ form a maximal r-cross-intersecting pair. By the previous paragraph, this pair can be transformed into a monotone pair by repeated applications of the shift operations $\phi_{i,j}$. Clearly, these operations do not introduce new relevant coordinates. It remains to check that the shifting operations do not produce balls from non-balls, that is, if $A, B \subseteq S_p$ are maximal r-cross-intersecting families, and $A' = \phi_{i,j}(A)$ and $B' = \phi_{i,j}(B)$ are balls, then so are A and B. In fact, by Lemma 4 it is sufficient to prove that one of them is a ball.

We saw that A' and B' must also form a maximal r-cross-intersecting pair. Thus, by Lemma 4, there is a set of coordinates $T \subseteq [n]$, a vector $x_0 \in S_p$, and radii l and m satisfying $|T| = r + l + m$ and that A' and B' are the Hamming balls of radius l and m in coordinates T around the vector x_0. We can assume that $i \in T$, because otherwise $A = A'$ and we are done. We also have that $(x_0)_i = 1$, as otherwise $A' = \phi_{i,j}(A)$ is impossible. The vectors $x \in S_p$ such that $x_i = j$ and

$$|\{k \in T : x_k \neq (x_0)_k\}| = l + 1$$

are called *A-critical*. Analogously, the vectors $y \in S_p$ such that $y_i = j$ and

$$|\{k \in T : y_k \neq (x_0)_k\}| = m + 1$$

are said to be B-*critical*. By the definition of $\phi_{i,j}$, the family A differs from A' by including some A-critical vectors x and losing the corresponding vectors $\phi_i(x)$. Symmetrically, $B \backslash B'$ consists of some B-critical vectors y and $B' \backslash B$ consists of the corresponding vectors $\phi_i(y)$. Let us consider the bipartite graph G whose vertices on one side are the A-critical vectors x, the vertices on the other side are the B-critical vectors y (considered as disjoint families, even if $l = m$), and x is adjacent to y if and only if $|\{k \in [n] : x_k = y_k\}| = r$. If x and y are adjacent, then neither the pair $(x, \phi_i(y))$, nor the pair $(\phi_i(x), y)$ is r-intersecting. As A and B are r-cross-intersecting, for any pair of adjacent vertices x and y of G, we have $x \in A$ if and only if $y \in B$.

The crucial observation is that the graph G is connected. Note that this is not the case if $p_k = 2$ for some index $k \notin T$, since all A-critical vectors x in a connected component of G would have the same value x_k. However, we assumed that $p_k > 2$ for $l \in [n]$. In this case, the A-critical vectors x and x' have a common B-critical neighbor (and, therefore, their distance in G is 2) if and only if the symmetric difference of the l element sets $\{k \in T \backslash \{i\} : x_k \neq (x_0)_k\}$ and $\{k \in T \backslash \{i\} : x'_k \neq (x_0)_k\}$ have at most $2r - 2$ elements. We assumed that $r > 1$, so this means that all A-critical vectors are indeed in the same component of the graph G. Therefore, either all A-critical vectors belong to A or none of them does. In the latter case, we have $A = A'$. In the former case, A is the Hamming ball of radius l in coordinates T around the vector x'_0, where x'_0 agrees with x_0 in all coordinates but in $(x'_0)_i = j$. In either case, A is a ball as required.   ⊔

Proof of Theorem 8. By Lemma 10, it is enough to restrict our attention to monotone families A and B. We may also assume that all coordinates are relevant (simply drop the irrelevant coordinates), and thus we have $n \leq r + 2$.

We denote by U_l the Hamming ball of radius l around the all-1 vector in the entire set of coordinates $[n]$. Notice that the monotone families A and B are r-cross-intersecting if and only if for $a \in \text{supp}(A)$ and $b \in \text{supp}(B)$ we have $|a \cup b| \leq n - r$. We consider all possible values of $n - r$, separately.

If $n = r$, both families A and B must coincide with the singleton U_0.

If $n = r + 1$, it is still true that either A or B is U_0. Otherwise, both $\text{supp}(A)$ and $\text{supp}(B)$ have to contain at least one non-empty set, but the union of these sets has size at most $n - r = 1$, so we have $\text{supp}(A) = \text{supp}(B) = \{\emptyset, \{i\}\}$ for some $i \in [n]$. But this contradicts our assumption that the coordinate i is relevant for A.

If $n = r + 2$, we are done if $A = B = U_1$. Otherwise, we must have a two-element set either in $\text{supp}(A)$ or in $\text{supp}(B)$. Let us assume that a two-element set $\{i, j\}$ belongs to $\text{supp}(A)$. Then each set $b \in \text{supp}(B)$ must satisfy $b \subseteq \{i, j\}$. This leaves five possibilities for a non-empty monotone family B, as $\text{supp}(B)$ must be one of the following set systems:

1. $\{\emptyset\}$,
2. $\{\emptyset, \{i\}\}$,
3. $\{\emptyset, \{j\}\}$,
4. $\{\emptyset, \{i\}, \{j\}\}$, and
5. $\{\emptyset, \{i\}, \{j\}, \{i, j\}\}$.

Cases 2, 3, and 5 are not possible, because either i or j would not be relevant for B.

In case 1, we have $B = U_0$, and thus $A = U_2$. In this case, A and B are balls, but the radius of A is 2. This is impossible, as U_1 is r-intersecting and $|U_1|^2 > |U_0| \cdot |U_2|$ always holds, so (A, B) is not maximal.

It remains to deal with case 4. Here $\mathrm{supp}(A)$ consists of the sets of size at most 1 and the two-element set $\{i, j\}$. Define

$$C = \{x \in S_p : x_k = 1 \text{ for all } k \in [n] \backslash \{i, j\}\}.$$

Note that $|A| + |B| = |U_1| + |C|$, because each vector in S_p appears in the same number of sets on both sides. Thus, we have either $|A| + |B| \leq 2|U_1|$ or $|A| + |B| \leq 2|C|$. Since $|A| > |B|$, the above inequalities imply $|A| \cdot |B| < |U_1|^2$ or $|A| \cdot |B| < |C|^2$. This contradicts the maximality of the pair (A, B), because both U_1 and C are r-intersecting. The contradiction completes the proof of Theorem 8. □

Let us remark here that the extension of Theorem 8 to somewhat larger values of relevant coordinates (in other words, verifying Conjecture 3 for the case of, say $r + 4$ relevant coordinates) yield a similar case analysis as we saw above for $r + 2$ relevant coordinates, but with much more cases corresponding to containment maximal pairs of set systems (U, V) with $|u \cup v|$ bounded for $u \in U$ and $v \in V$. This seems to be doable, but the number of cases to considers grow fast.

Now we can prove our main theorem, verifying Conjecture 3 in several special cases.

Proof of Theorem 7. The statement about the case $p_1 > r + 1$ readily follows from Lemma 2, as in this case the condition

$$\frac{2}{p_{r+1}} + \sum_{i=1}^{r} \frac{1}{p_i} \leq 1$$

holds.

To prove the other two statements in the theorem, we assume that A and B form a maximal r-cross-intersecting pair. We also assume without loss of generality that all coordinates are relevant for both families (simply drop the irrelevant coordinates).

By Theorem 6, we have $\prod_{i=1}^{r} p_i \geq \prod_{i=1}^{n} (p_i - 1)^{1-2/p_i}$, and thus

$$\prod_{i=1}^{r} \frac{p_i}{(p_i - 1)^{1-2/p_i}} \geq \prod_{i=r+1}^{n} (p_i - 1)^{1-2/p_i}.$$

Here the function $x/(x - 1)^{1-2/x}$ is decreasing for $x \geq 3$, while $(x - 1)^{1-2/x}$ is increasing, and we have $p_i \geq p_1 \geq 3$. Therefore, we also have

$$\prod_{i=1}^{r} \frac{p_1}{(p_1 - 1)^{1-2/p_1}} \geq \prod_{i=r+1}^{n} (p_1 - 1)^{1-2/p_1},$$

$$p_1^r \geq (p_1 - 1)^{n(1 - 2/p_1)},$$

$$n \leq \frac{r \log p_1}{(1 - 2/p_1) \log(p_1 - 1)}.$$

It can be shown by simple computation that the right-hand side of the last inequality is strictly smaller than $r + 3$ if $p_1 \leq 2r/3 + t(r)$ for some function $t(r) = O(r/\log r)$ and, in particular, for $p_1 = r + 1 \geq 8$. In this case, we have $n \leq r + 2$ relevant coordinates. Thus, Theorem 8 applies, yielding that A and B are balls. This proves the last statement of Theorem 7.

For the proof of the second statement, note that we have already established that A and B are balls of radius 0 or 1. We use Lemma 4 to calculate the sizes of A in B in the three possible cases. The product $|A| \cdot |B|$ is $z_1 = \prod_{i=r+1}^{n} p_i^2$ if A and B are balls of radius 0. The same product is $z_2 = (\sum_{i=1}^{r+1} p_i - r) \prod_{i=r+2}^{n} p_i^2$ if one of them is a ball of radius 0 while the other is a ball of radius 1. Finally, the product is $z_3 = (\sum_{i=1}^{r+2} p_i - r - 1)^2 \prod_{i=r+3}^{n} p_i^2$ if both families are balls of radius 1. Note that we have $A = B$ in the first and third cases. Using the condition $p_i \geq r + 1$, it is easy to verify that $z_2 < z_1$ and $z_3 \leq z_1$. Furthermore, we have $z_3 = z_1$ if and only if $p_i = r + 1$ for all $i \in [r + 2]$. This completes the proof of Theorem 7.   □

5   Coordinates with $p_i = 2$

In many of our results, we had to assume $p_i > 2$ for all coordinates of the size vector. Here we elaborate on why the coordinates $p_i = 2$ behave differently.

For the simple characterization of the cases of equality in Theorem 1, the assumption $k \neq 2$ is necessary. Here we characterize all maximal cross-intersecting pairs in the case $k = 2$.

Let $p = (p_1, \ldots, p_n)$ be a size vector of positive integers with $k = \min_i p_i = 2$ and let $I = \{i \in [n] : p_i = 2\}$. For any set W of functions $I \to [2]$, define the families

$$A_W = \{x \in S_p : \exists f \in W \text{ such that } x_i = f(i) \text{ for every } i \in I\},$$

$$B_W = \{y \in S_p : \nexists f \in W \text{ such that } y_i \neq f(i) \text{ for every } i \in I\}.$$

The families A_W and B_W are cross-intersecting for any W. Moreover, if $|W| = 2^{|I|-1}$, we have $|A_W| \cdot |B_W| = |S_p|^2/4$, so they form a maximal cross-intersecting pair. Note that these include more examples than just the pairs of families described in Theorem 1, provided that $|I| > 1$.

We claim that all maximal cross-intersecting pairs are of the form constructed above. To see this, consider a maximal pair $A, B \subseteq S_p$. We know from the proof of Theorem 1 that $x \in A$ if and only if $x' \notin B$, where x' is defined by $x_i' = (x_i + 1 \bmod p_i)$ for all $i \in [n]$. Let $j \in [n]$ be a coordinate with $p_j > 2$. By the same argument, we also have that $x \in A$ holds if and only if $x'' \notin B$, where $x_i'' = x_i'$ for $i \in [n] \backslash \{j\}$ and $x_j'' = (x_j + 2 \bmod p_j)$. Thus, both x' and x'' belong to B or neither of them does. This holds for every vector x', implying that j is irrelevant for the family B and thus also for A.

As there are no relevant coordinates for A and B outside the set I of coordinates with $p_i = 2$, we can choose a set W of functions from I to $[2]$ such that $A = A_W$. This makes

$$B = \{y \in S_p \ : \ y \text{ intersects all } x \in A\} = B_W.$$

We have $|A| + |B| = |S_p|$ and $|A| \cdot |B| = |S_p|^2/4$ if and only if $|W| = 2^{|I|-1}$.

The size vector $p = (2, \ldots, 2)$ of length n is well studied. In this case, S_p is the n-dimensional hypercube. If $r > 1$, then all maximal r-cross-intersecting pairs have an unbounded number of relevant coordinates, as a function of n. Indeed, the density $|A| \cdot |B|/|S_p|^2$ is at most $1/4$ for cross-intersecting pairs $A, B \subseteq S_p$, and strictly less than $1/4$ for r-cross-intersecting families if $r > 1$. Furthermore, if the number of relevant coordinates is bounded, then this density is bounded away from $1/4$, while if $A = B$ is the ball of radius $(n-r)/2$ in all the coordinates, then the same density approaches $1/4$.

One can also find many maximal 2-cross-intersecting pairs that are not balls. For example, in the 3-dimensional hypercube the families $A = \{0,0,0), (0,1,1)\}$ and $B = \{(0,0,1), (0,1,0)\}$ form a maximal 2-cross-intersecting pair.

Finally, we mention that there is a simple connection between the problem discussed in this paper and a question related to communication complexity. Consider the following two-person communication game: Alice and Bob each receive a vector from S_p, and they have to decide whether the vectors are r-intersecting. In the *communication matrix* of such a game, the rows are indexed by the possible inputs of Alice, the columns by the possible inputs of Bob, and an entry of the matrix is 1 or 0 corresponding to the "yes" or "no" output the players have to compute for the corresponding inputs. In the study of communication games, the submatrices of this matrix in which all entries are equal play a special role. The largest area of an all-1 submatrix is the maximal value of $|A| \cdot |B|$ for r-cross-intersecting families $A, B \subseteq S_p$.

Acknowledgment. We are indebted to G. O. H. Katona, R. Radoičić, and D. Scheder for their valuable remarks, and to an anonymous referee for calling our attention to the manuscript of Borg [Bo14].

References

[AhK98] Ahlswede, R., Khachatrian, L.H.: A diametric theorem for edges. The diametric theorem in Hamming spaces—optimal anticodes. Adv. Appl. Math. **20**(4), 429–449 (1998)

[Be74] Berge, C.: Nombres de coloration de l'hypegraphe h-parti complet. In: Hypergraph Seminar. Lecture Notes in Mathematics, vol. 411, pp. 13–20. Springer, Heidelberg (1974)

[Bol86] Bollobás, B.: Combinatorics: Set Systems, Hypergraphs, Families of Vectors and Combinatorial Probability. Cambridge University Press, Cambridge (1986)

[Bo08] Borg, P.: Intersecting and cross-intersecting families of labeled sets. Electron. J. Combin. **15**, N9 (2008)

[Bo14] Borg, P.: Cross-intersecting integer sequences. Preprint. arXiv:1212.6955

[DeF83] Deza, M., Frankl, P.: Erdős-Ko-Rado theorem – 22 years later. SIAM J. Algebr. Discrete Methods **4**, 419–431 (1983)

[EKR61] Erdős, P., Ko, C., Rado, R.: Intersection theorems for systems of finite sets. Q. J. Math. Oxford Ser. **2**(12), 313–318 (1961)

[F95] Frankl, P.: Extremal set systems. In: Graham, R., et al. (eds.) Handbook of Combinatorics, pp. 1293–1329. Elsevier, Amsterdam (1995)

[FF80] Frankl, P., Füredi, Z.: The Erdős-Ko-Rado theorem for integer sequences. SIAM J. Algebr. Discrete Methods **1**, 376–381 (1980)

[FLST14] Frankl, P., Lee, S.J., Siggers, M., Tokushige, N.: An Erdős-Ko-Rado theorem for cross-intersecting families. Preprint. arXiv:1303.0657

[FT99] Frankl, P., Tokushige, N.: The Erdős-Ko-Rado theorem for integer sequences. Combinatorica **19**, 55–63 (1999)

[GK78] Greene, C., Kleitman, D.J.: Proof techniques in the ordered sets. In: Studies in Combinatorics, pp. 22–79. Mathematical Association of America, Washington, DC (1978)

[Hi77] Hilton, A.J.W.: An intersection theorem for a collection of families of subsets of a finite set. J. Lond. Math. Soc. **2**, 369–384 (1977)

[K72] Katona, G.O.H.: A simple proof of the Erdős-Ko-Rado theorem. J. Combin. Theory Ser. B **13**, 183–184 (1972)

[Liv79] Livingston, M.L.: An ordered version of the Erdős-Ko-Rado theorem. J. Combin. Theory Ser. A **26**, 162–165 (1979)

[Mo82] Moon, A.: An analogue of the Erdős-Ko-Rado theorem for the Hamming schemes H(n, q). J. Combin. Theory Ser. A **32**(3), 386–390 (1982)

[Py86] Pyber, L.: A new generalization of the Erdős-Ko-Rado theorem. J. Combin. Theory Ser. A **43**, 85–90 (1986)

[St80] Stanton, D.: Some Erdős-Ko-Rado theorems for Chevalley groups. SIAM J. Algebr. Discrete Methods **1**(2), 160–163 (1980)

[To13] Tokushige, N.: Cross t-intersecting integer sequences from weighted Erdős-Ko-Rado. Combin. Prob. Comput. **22**, 622–637 (2013)

[Zh13] Zhang, H.: Cross-intersecting families of labeled sets. Electron. J. Combin. **20**(1), P17 (2013)

The Double Multicompetition Number of a Multigraph

Jeongmi Park[1] and Yoshio Sano[2]([⊠])

[1] Department of Mathematics, Pusan National University, Busan 609-735, Korea
jm1015@pusan.ac.kr
[2] Division of Information Engineering, Faculty of Engineering,
Information and Systems, University of Tsukuba, Ibaraki 305-8573, Japan
sano@cs.tsukuba.ac.jp

Abstract. The double competition multigraph of a digraph D is the multigraph which has the same vertex set as D and has m_{xy} multiple edges between two distinct vertices x and y, where m_{xy} is defined to be the number of common out-neighbors of x and y in D times the number of common in-neighbors of x and y in D.

In this paper, we introduce the notion of the double multicompetition number of a multigraph. It is easy to observe that, for any multigraph M, M together with sufficiently many isolated vertices is the double competition multigraph of some acyclic digraph. The double multicompetition number of a multigraph is defined to be the minimum number of such isolated vertices. We give a characterization of multigraphs with bounded double multicompetition number and give a lower bound for the double multicompetition numbers of multigraphs.

Keywords: Intersection graph · Double competition multigraph · Double multicompetition number · Acyclic digraph

2010 Mathematics Subject Classification: 05C20, 05C75.

1 Introduction

The competition graph of a digraph is defined to be the intersection graph of the family of the out-neighborhoods of the vertices of the digraph (see [11] for intersection graphs). A *digraph* D is a pair $(V(D), A(D))$ of a set $V(D)$ of *vertices* and a set $A(D)$ of ordered pairs of vertices, called *arcs*. An arc of the form (v, v) is called a *loop*. For a vertex x in a digraph D, we denote the *out-neighborhood* of x in D by $N_D^+(x)$ and the *in-neighborhood* of x in D by $N_D^-(x)$, i.e., $N_D^+(x) := \{v \in V(D) \mid (x, v) \in A(D)\}$ and $N_D^-(x) := \{v \in V(D) \mid (v, x) \in A(D)\}$. A *graph* G is a pair $(V(G), E(G))$ of a set $V(G)$ of *vertices* and a set $E(G)$ of unordered pairs of vertices, called *edges*. The *competition graph* of a digraph D is the graph

Yoshio Sano - This work was supported by JSPS KAKENHI grant number 25887007.

© Springer International Publishing Switzerland 2014
J. Akiyama et al. (Eds.): JCDCGG 2013, LNCS 8845, pp. 138–144, 2014.
DOI: 10.1007/978-3-319-13287-7_12

which has the same vertex set as D and has an edge between two distinct vertices x and y if and only if $N_D^+(x) \cap N_D^+(y) \neq \emptyset$. R. D. Dutton and R. C. Brigham [4] and F. S. Roberts and J. E. Steif [13] gave characterizations of competition graphs by using edge clique covers of graphs. The notion of competition graphs was introduced by J. E. Cohen [3] in 1968 in connection with a problem in ecology, and several variants and generalizations of competition graphs have been studied.

In 1987, D. D. Scott [16] introduced the notion of double competition graphs as a variant of the notion of competition graphs. The *double competition graph* (or the *competition-common enemy graph* or the *CCE graph*) of a digraph D is the graph which has the same vertex set as D and has an edge between two distinct vertices x and y if and only if both $N_D^+(x) \cap N_D^+(y) \neq \emptyset$ and $N_D^-(x) \cap N_D^-(y) \neq \emptyset$ hold. See [8,10,15] for recent results on double competition graphs.

A *multigraph* M is a pair $(V(M), E(M))$ of a set $V(M)$ of *vertices* and a multiset $E(M)$ of unordered pairs of vertices, called *edges*. Note that, in our definition, multigraphs have no loops. We may consider a multigraph M as the pair $(V(M), m_M)$ of the vertex set $V(M)$ and the nonnegative integer-valued function $m_M : \binom{V}{2} \to \mathbb{Z}_{>0}$ on the set $\binom{V}{2}$ of all unordered pairs of V where $m_M(\{x, y\})$ is defined to be the number of multiple edges between the vertices x and y in M. The notion of competition multigraphs was introduced by C. A. Anderson, K. F. Jones, J. R. Lundgren, and T. A. McKee [1] in 1990 as a variant of the notion of competition graphs. The *competition multigraph* of a digraph D is the multigraph which has the same vertex set as D and has m_{xy} multiple edges between two distinct vertices x and y, where m_{xy} is the nonnegative integer defined by $m_{xy} = |N_D^+(x) \cap N_D^+(y)|$. See [14,19] for recent results on competition multigraphs.

In [12], the authors introduced the notion of the double competition multigraph of a digraph. The *double competition multigraph* of a digraph D is the multigraph which has the same vertex set as D and has m_{xy} multiple edges between two distinct vertices x and y, where m_{xy} is the nonnegative integer defined by

$$m_{xy} = |N_D^+(x) \cap N_D^+(y)| \cdot |N_D^-(x) \cap N_D^-(y)|.$$

Recall that a *clique* of a multigraph M is a set of vertices of M which are pairwise adjacent. We consider the empty set $\emptyset$ as a clique of any multigraph for convenience. A multiset is also called a *family*. An *edge clique partition* of a multigraph M is a family $\mathcal{F}$ of cliques of M such that any two distinct vertices x and y are contained in exactly $m_M(\{x, y\})$ cliques in the family $\mathcal{F}$.

A digraph D is said to be *acyclic* if D has no directed cycles. An ordering $(v_1, \ldots, v_n)$ of the vertices of a digraph D, where n is the number of vertices of D, is called an *acyclic ordering* of D if $(v_i, v_j) \in A(D)$ implies $i < j$. It is well known that a digraph D is acyclic if and only if D has an acyclic ordering. For a positive integer n, let $[n]$ denote the set $\{1, 2, \ldots, n\}$.

In [12], the authors gave a characterization of the double competition multigraphs of acyclic digraphs in terms of edge clique partitions of the multigraphs.

Theorem 1 [12]. *Let M be a multigraph with n vertices. Then, M is the double competition multigraph of an acyclic digraph if and only if there exist an ordering $(v_1, \ldots, v_n)$ of the vertices of M and a double indexed edge clique partition $\{S_{ij} \mid i, j \in [n]\}$ of M such that the following conditions hold:*

(I) *for any $i, j \in [n]$, if $|A_i \cap B_j| \geq 2$, then $A_i \cap B_j = S_{ij}$;*
(IV) *for any $i, j, k \in [n]$, $v_k \in S_{ij}$ implies $i < k < j$,*

where A_i and B_j are the sets defined by

$$A_i = S_{i*} \cup T_i^+, \quad S_{i*} := \bigcup_{p \in [n]} S_{ip}, \quad T_i^+ := \{v_b \mid a, b \in [n], v_i \in S_{ab}\},$$

$$B_j = S_{*j} \cup T_j^-, \quad S_{*j} := \bigcup_{q \in [n]} S_{qj}, \quad T_j^- := \{v_a \mid a, b \in [n], v_j \in S_{ab}\}.$$

In this paper, we introduce the notion of the double multicompetition number of a multigraph, and we give a characterization of multigraphs with bounded double multicompetition number and give a lower bound for the double multicompetition numbers of multigraphs.

2   Main Results

Proposition 2. *For any multigraph M, there exists a nonnegative integer k such that $M \cup I_k$ is the double competition multigraph of an acyclic digraph, where I_k is the set of k isolated vertices.*

Proof. Let $M = (V(M), m_M)$ be a multigraph. Let $E_1 := \{\{x, y\} \in \binom{V(M)}{2} \mid m_M(\{x, y\}) \geq 1\}$, let $A := \{a_{\{x,y\}} \mid \{x, y\} \in E_1\}$, and let $Z := \{z_{\{x,y\},i} \mid \{x, y\} \in E_1, i \in \{1, \ldots, m_M(\{x, y\})\}\}$. We define a digraph D by

$$V(D) := V(M) \cup A \cup Z$$

$$A(D) := \bigcup_{\{x,y\} \in E_1} \bigcup_{i=1}^{m_M(\{x,y\})} \{(a_{\{x,y\}}, x), (a_{\{x,y\}}, y), (x, z_{\{x,y\},i}), (y, z_{\{x,y\},i})\}.$$

Then, the digraph D is acyclic and the double competition multigraph of D is $M \cup I_k$, where $k = |A| + |Z|$. Hence the proposition holds.   □

By Proposition 2, we can define the following:

Definition. For a multigraph M, the *double multicompetition number* of M is the minimum nonnegative integer k such that $M \cup I_k$ is the double competition multigraph of an acyclic digraph. We denote it by $dk^*(M)$.   □

Lemma 3. *If a multigraph M has no isolated vertices, then $dk^*(M) \geq 2$.*

Proof. Let $k := dk^*(M)$. Then, there exists an acyclic digraph D whose double competition multigraph is $M \cup I_k$. Let $(v_1, \ldots, v_{n+k})$ be an acyclic ordering of D. Let i be the minimum index such that $v_i \in V(M)$ and let j be the maximum index such that $v_j \in V(M)$. Then $k \geq (i-1)+(n+k-j)$. Since M has no isolated vertices, the vertex v_i (and v_j) has at least one incident edge. Therefore, v_i (and v_j) must have both an in-neighbor and an out-neighbor in D. Since v_i has an in-neighbor, we have $i \geq 2$. Since v_j has an out-neighbor, we have $j \leq n+k-1$. Thus we obtain $k \geq 2$. Hence the lemma holds. $\quad\square$

Example 4. *For a path P_n of order n with $n \geq 2$, $dk^*(P_n) = 2$.*

Proof. Let P_n be the path with $V(P_n) = \{v_1, \ldots, v_n\}$ and $E(P_n) = \{\{v_i, v_{i+1}\} \mid i \in \{1, \ldots, n-1\}\}$. We define a digraph D by $V(D) = \{v_1, \ldots, v_n\} \cup \{a, z\}$ and $A(D) = \{(a, v_1), (a, v_2)\} \cup \left(\bigcup_{i=1}^{n-2}\{(v_i, v_{i+1}), (v_i, v_{i+2})\}\right) \cup \{(v_{n-1}, v_n)\} \cup \{(v_{n-1}, z), (v_n, z)\}$, where a and z are new vertices. Then, the digraph D is acyclic since the ordering $(a, v_1, \ldots, v_n, z)$ is an acyclic ordering of D. We can easily check that the double competition multigraph of D is $P_n \cup \{a, z\}$. Thus $dk^*(P_n) \leq 2$. By Lemma 3, $dk^*(P_n) \geq 2$. Hence $dk^*(P_n) = 2$. $\quad\square$

Now, we give a characterization for multigraphs with bounded double multicompetition number, which is an extension of Theorem 1.

Theorem 5. *Let M be a multigraph with n vertices. Let k be a nonnegative integer. Then, the double multicompetition number of M is at most k if and only if there exist an integer t such that $0 \leq t \leq k$, an ordering $(v_1, \ldots, v_n)$ of the vertices of M, and a double indexed edge clique partition $\{S_{ij} \mid i, j \in [n+k]\}$ of M such that the following conditions hold:*

(i) *For any $i, j \in [n]$, if $|A_i \cap B_j| \geq 2$, then $A_i \cap B_j = S_{t+i,t+j}$, where A_i and B_j are the sets defined for $i, j \in [n]$ by*

$$A_i = \left(\bigcup_{p \in [n+k-t]} S_{t+i,t+p}\right) \cup \{v_{b-t} \mid v_i \in S_{ab} \ (a, b \in \{t+1, \ldots, t+n\})\},$$

$$B_j = \left(\bigcup_{q \in [n+k-t]} S_{t+q,t+j}\right) \cup \{v_{a-t} \mid v_j \in S_{ab} \ (a, b \in \{t+1, \ldots, t+n\})\}.$$

(ii) *Let $\Lambda_1 = \{1, \ldots, t\}$, $\Lambda_2 = \{t+1, \ldots, t+n\}$, and $\Lambda_3 = \{t+n+1, \ldots, t+n+(k-t)(= n+k)\}$.*
If $(i, j) \in \Lambda_1 \times \Lambda_2$, then $S_{ij} \subseteq \{v_1, \ldots, v_{(j-t)-1}\}$.
If $(i, j) \in \Lambda_1 \times \Lambda_3$, then $S_{ij} \subseteq \{v_1, \ldots, v_n\}$.
If $(i, j) \in \Lambda_2 \times \Lambda_2$ and $i < j$, then $S_{ij} \subseteq \{v_{(i-t)+1}, \ldots, v_{(j-t)-1}\}$.
If $(i, j) \in \Lambda_2 \times \Lambda_3$, then $S_{ij} \subseteq \{v_{(i-t)+1}, \ldots, v_n\}$.

Otherwise, $S_{ij} = \emptyset$.

Proof. First, we show the "only-if" part. Suppose that a multigraph M has the double multicompetition number at most k. Let $M' := M \cup I_k$. Then M' is the double competition multigraph of an acyclic digraph. By Theorem 1, there exist an ordering $(v'_1, \ldots, v'_{n+k})$ of the vertices of M' and a double indexed edge clique partition $\mathcal{F} = \{S_{ij} \mid i, j \in [n+k]\}$ of M' such that the conditions (I) and (IV) in Theorem 1 hold. Without loss of generality, we may assume that $I_k = \{v'_1, \ldots, v'_t\} \cup \{v'_{n+t+1}, \ldots, v'_{n+k}\}$ for some nonnegative integer t with $0 \leq t \leq k$. We delete all the cliques of size one from $\mathcal{F}$ if they exist. Since the vertices in I_k are isolated, there is no clique S_{ij} of size at least two containing a vertex in I_k. Let $(v_1, \ldots, v_n) := (v'_{t+1}, \ldots, v'_{t+n})$. Then, it follows from the conditions (I) and (IV) for $(v'_1, \ldots, v'_{n+k})$ and $\mathcal{F}$ that the integer t, the ordering $(v_1, \ldots, v_n)$, and the family $\mathcal{F}$ satisfy the conditions (i) and (ii).

Next, we show the "if" part. Suppose that there exist an integer t such that $0 \leq t \leq k$, an ordering $(v_1, \ldots, v_n)$ of the vertices of M, and a double indexed edge clique partition $\{S_{ij} \mid i, j \in [n+k]\}$ of M such that the conditions (i) and (ii) hold. Then, we define a digraph D by $V(D) := V(M) \cup A \cup Z$, where $A := \{a_1, \ldots, a_t\}$ and $Z := \{z_1, \ldots, z_{k-t}\}$, and

$$A(D) := \bigcup_{(i,j)\in\Lambda_1\times\Lambda_2} \left(\bigcup_{v\in S_{ij}} \{(a_i, v), (v, v_{j-t})\} \right)$$
$$\cup \bigcup_{(i,j)\in\Lambda_1\times\Lambda_3} \left(\bigcup_{v\in S_{ij}} \{(a_i, v), (v, z_{j-(n+t)})\} \right)$$
$$\cup \bigcup_{(i,j)\in\Lambda_2\times\Lambda_2, i<j} \left(\bigcup_{v\in S_{ij}} \{(v_{i-t}, v), (v, v_{j-t})\} \right)$$
$$\cup \bigcup_{(i,j)\in\Lambda_2\times\Lambda_3} \left(\bigcup_{v\in S_{ij}} \{(v_{i-t}, v), (v, z_{j-(n+t)})\} \right).$$

Then, we can check that the ordering $(a_1, \ldots, a_t, v_1, \ldots, v_n, z_1, \ldots, z_{k-t})$ is an acyclic ordering of D. Therefore the digraph D is acyclic. By the definition of the digraph D, it follows that

$$N_D^+(v_h) = \bigcup_{j\in\Lambda_2\cup\Lambda_3, t+h<j} S_{t+h,j}$$
$$\cup \{v_{j-t} \mid v_h \in S_{ij}, (i,j) \in (\Lambda_1 \cup \Lambda_2) \times \Lambda_2, i < j\}$$
$$\cup \{z_{j-(n+t)} \mid v_h \in S_{ij}, (i,j) \in (\Lambda_1 \cup \Lambda_2) \times \Lambda_3\},$$
$$N_D^-(v_h) = \bigcup_{i\in\Lambda_1\cup\Lambda_2, i<t+h} S_{i,t+h}$$
$$\cup \{a_i \mid v_h \in S_{ij}, (i,j) \in \Lambda_1 \times (\Lambda_2 \cup \Lambda_3)\}$$
$$\cup \{v_{i-t} \mid v_h \in S_{ij}, (i,j) \in \Lambda_2 \times (\Lambda_2 \cup \Lambda_3), i < j\}$$

for $h \in [n]$, and that

$$N_D^+(a_i) = \bigcup_{j \in \Lambda_2 \cup \Lambda_3} S_{ij}, \qquad N_D^-(a_i) = \emptyset \qquad \text{for } i \in [t],$$

$$N_D^-(z_j) = \bigcup_{i \in \Lambda_1 \cup \Lambda_2} S_{i,t+n+j}, \qquad N_D^+(z_j) = \emptyset \qquad \text{for } j \in [k-t].$$

Therefore, we can confirm that the double competition multigraph of D is the multigraph $M \cup (A \cup Z)$, where the vertices in $A \cup Z$ are isolated. Thus, we have $dk^*(M) \leq |A| + |Z| = k$.

Hence the theorem holds. $\qquad\qquad\qquad\qquad\qquad\qquad\qquad\qquad\qquad\qquad\qquad\qquad$ □

Let $\theta_E^*(M)$ denote the *edge clique partition number* of a multigraph M, i.e., the minimum size of an edge clique partition of M. By using Theorem 5, we obtain a lower bound for $dk^*(M)$.

Theorem 6. *For any multigraph M with n vertices,*

$$dk^*(M) \geq \max \left\{ 0, \left\lceil \sqrt{4\theta_E^*(M) + 2n^2 + 2n} \right\rceil - 2n \right\}.$$

Proof. Let $k := dk^*(M)$. Then, by Theorem 5, there exist an ordering $(v_1, \ldots, v_n)$ of the vertices of M and a double indexed edge clique partition $\mathcal{F} = \{S_{ij} \mid i, j \in [n]\}$ of M such that the conditions (i) and (ii) hold. Then, it follows from condition (ii) that the number of nonempty cliques in the family $\mathcal{F}$ is at most

$$|\Lambda_1 \times \Lambda_2| + |\Lambda_1 \times \Lambda_3| + |(\Lambda_2 \times \Lambda_2) \cap \{(i,j) \mid i < j\}| + |\Lambda_2 \times \Lambda_3|$$

$$= tn + t(k-t) + \frac{1}{2}n(n-1) + n(k-t)$$

$$= -\left(t - \frac{1}{2}k\right)^2 + \frac{1}{4}k^2 + nk + \frac{1}{2}n(n-1).$$

Therefore, we have an inequality $\theta_E^*(M) \leq \frac{1}{4}k^2 + nk + \frac{1}{2}n(n-1)$, that is,

$$k^2 + 4nk + 2n(n-1) - 4\theta_E^*(M) \geq 0.$$

Since $k \geq 0$, we obtain $k \geq \max\{0, \sqrt{4\theta_E^*(M) + 2n^2 + 2n} - 2n\}$. Since k is an integer, the theorem holds. $\qquad\qquad\qquad\qquad\qquad\qquad\qquad\qquad\qquad\qquad\qquad$ □

The *double competition number* $dk(G)$ of a graph G is defined to be the minimum nonnegative integer k such that $G \cup I_k$ is the double competition graph of an acyclic digraph. See [2,5–7,9,16–18] for results on the double competition numbers of graphs. Note that a graph can be considered as a multigraph without multiple edges. The following gives a relationship between the double competition number $dk(G)$ and the double multicompetition number $dk^*(G)$ of a graph.

Proposition 7. *For any graph G, $dk(G) \leq dk^*(G)$.*

Proof. Let D be an acyclic digraph such that the double competition multigraph of D is the (multi)graph $G \cup I_{dk^*(G)}$. Then, the double competition graph of D is also the graph $G \cup I_{dk^*(G)}$. Thus, we have $dk(G) \leq dk^*(G)$. $\qquad\qquad\qquad$ □

References

1. Anderson, C.A., Jones, K.F., Lundgren, J.R., McKee, T.A.: Competition multi-graphs and the multicompetition number. Ars Combin. **29B**, 185–192 (1990)
2. Bak, O.-B., Kim, S.-R.: On the double competition number of a bipartite graph. Congress. Numeran. **117**, 145–152 (1996)
3. Cohen, J.E.: Interval graphs and food webs. A finding and a problem. RAND Corporation Document 17696-PR, Santa Monica, CA (1968)
4. Dutton, R.D., Brigham, R.C.: A characterization of competition graphs. Discrete Appl. Math. **6**, 315–317 (1983)
5. Füredi, Z.: On the double competition number. Discrete Appl. Math. **82**, 251–255 (1998)
6. Jones, F.K., Lundgren, J.R., Roberts, F.S., Seager, S.: Some remarks on the double competition number of a graph. Congress. Numeran. **60**, 17–24 (1987)
7. Kim, S.-R.: On the inequality $dk(G) \leq k(G) + 1$. Ars Combin. **51**, 173–182 (1999)
8. Kim, S.-J., Kim, S.-R., Rho, Y.: On CCE graphs of doubly partial orders. Discrete Appl. Math. **155**, 971–978 (2007)
9. Kim, S.-R., Roberts, F.S., Seager, S.: On 1 0 1-clear $(0,1)$ matrices and the double competition number of bipartite graphs. J. Combin. Inf. Syst. Sci. **17**, 302–315 (1992)
10. Lu, J., Wu, Y.: Two minimal forbidden subgraphs for double competition graphs of posets of dimension at most two. Appl. Math. Lett. **22**, 841–845 (2009)
11. McKee, T.A., McMorris, F.R.: Topics in Intersection Graph Theory. SIAM Monographs on Discrete Mathematics and Applications. Society for Industrial and Applied Mathematics (SIAM), Philadelphia (1999)
12. Park, J., Sano, Y.: The double competition multigraph of a digraph. Preprint, July 2013. arXiv:1307.5509
13. Roberts, F.S., Steif, J.E.: A characterization of competition graphs of arbitrary digraphs. Discrete Appl. Math. **6**, 323–326 (1983)
14. Sano, Y.: Characterizations of competition multigraphs. Discrete Appl. Math. **157**, 2978–2982 (2009)
15. Sano, Y.: The competition-common enemy graphs of digraphs satisfying Conditions $C(p)$ and $C'(p)$. Congress. Numeran. **202**, 187–194 (2010)
16. Scott, D.D.: The competition-common enemy graph of a digraph. Discrete Appl. Math. **17**, 269–280 (1987)
17. Seager, S.M.: The double competition number of some triangle-free graphs. Discrete Appl. Math. **28**, 265–269 (1990)
18. Wu, Y., Lu, J.: Dimension-2 poset competition numbers and dimension-2 poset double competition numbers. Discrete Appl. Math. **158**, 706–717 (2010)
19. Zhao, Y., Chang, G.J.: Multicompetition numbers of some multigraphs. Ars Combin. **97**, 457–469 (2010)

Computational Geometry in the Human Brain

Kokichi Sugihara[✉]

Graduate School of Advanced Mathematical Sciences,
Meiji University, 4-21-1 Nakano, Nakano-ku, Tokyo 164-8525, Japan
kokichis@isc.meiji.ac.jp
http://home.mims.meiji.ac.jp/~sugihara

Abstract. Geometric information processing in the human brain is very different from that in a computer: it is slow, local, and imprecise. However, humans are able to manage a huge amount of visual data, can understand the scenes in front of them, and thus can survive in their daily lives. We use visual illusions to investigate how the human brain treats geometric data, and we point out the similarities between the robustness of human geometric processing and the topology-oriented principle, which we have proposed for use in the design of robust geometric algorithms for computers by presenting a new algorithm for straight skeletons.

Keywords: Visual illusion · Human vision · Robust geometric algorithm · Topology-oriented approach · Brain computing · Zöllner illusion · Ouchi illusion · Impossible motion · Straight skeleton

1 Introduction

Computational geometry is the field in which geometric algorithms are designed for computers [5,7,17]. Computers are much more precise than human brains, and hence the main concern is to make these algorithms as efficient as possible [1]. Indeed, a huge number of very efficient algorithms have been established, and sometimes these are the most efficient, i.e., optimal, in terms of the order of the computational time with respect to the problem size. In this sense, computational geometry is one of the most successful areas of computer science.

However, we note that computational geometry mainly treats well-defined problems, while in the real world, we encounter many geometric problems that are not well defined and cannot be solved easily by the current techniques of computational geometry. Such problems include, for example, image pattern recognition and scene understanding [3].

The human brain, on the other hand, seems to be able to solve those problems relatively easily. We receive geometric information about the world around us in the form of projected images on the retina, and our brains process those images and understand the scenes in front of us without any major difficulties. This ability is surprising when we recall that the computations of the human brain are slow and imprecise, compared to electronic computers. If we can understand

© Springer International Publishing Switzerland 2014
J. Akiyama et al. (Eds.): JCDCGG 2013, LNCS 8845, pp. 145–160, 2014.
DOI: 10.1007/978-3-319-13287-7_13

the way in which human brains perform geometric processing, we may be able to apply it to the construction of algorithms for ill-defined geometric problems.

Motivated by this observation, we used visual illusions to study the way in which the human brain processes geometric data, in order to find an approach for solving ill-defined problems in computational geometry.

On the other hand, our research group has long been studying an approach to the design of robust geometric algorithms, which we call the topology-oriented approach [22,24,26]. In this approach we start with the assumption that numerical errors cannot be avoided and moreover the amount of errors is not bounded a priori, but still we aim at robust geometric algorithms. This task is ill-conditioned because the correctness of the algorithms cannot be guaranteed due to numerical errors. However, we can successfully construct stable algorithms by guaranteeing the consistency of topological structures of geometric objects and thus circumvent failures.

We first applied this idea to an incremental algorithm for ordinary Voronoi diagrams [25], and then extended to various geometric problems including Voronoi diagrams for line segments [26] and line arrangements [10] in the plane, and convex hull [15], Voronoi diagram [26] and polyhedra [20] in the three-dimensional space.

Therefore, we might regard the topology-oriented approach as an example of human-like robust computation. In this paper we compare human brain processing with the topology-oriented algorithms and discuss their similarities.

The structure of this paper is as follows. We first observe and discuss three typical examples of visual illusions, the Zöllner illusion [11], the Ouchi illusion [14], and the impossible motion illusion [21], in Sects. 2–4, respectively. In Sect. 5, we construct a new algorithm for robust computation of the straight skeleton as an example of a geometric problem, and discuss the similarities between the computations in the human brain and the topology-oriented algorithms. We present our concluding remarks in Sect. 6.

2   Zöllner Illusion and Overestimation of Acute Angles

Figure 1 shows the famous Zöllner illusion; the four long, straight lines are exactly parallel and horizontal, but they look as if they are alternately slanted in opposite directions. This optical illusion is evoked by the shorter lines crossing the longer lines, and it is usually explained by the overestimation of the acute angles.

When two lines cross, they generate two acute angles and two obtuse angles. It is commonly observed that the acute angles are apt to be perceived larger than they actually are, and the obtuse angles are apt to be perceived smaller. There are many other illusions explained in the same way, including the Hering illusion, the Wundt illusion, and the Luckiesh illusion [11].

Various mathematical models have been proposed to explain this overestimation of acute angles. A typical such model is the one by Fremüller et al. [9]. In their theory, the retina, acting as a photo sensor, has finite resolution, and hence images are blurred, resulting in greater rounding of acute angles than of obtuse angles. This makes acute angles appear to be greater than the actual angles.

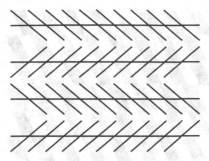

Fig. 1. Zöllner illusion.

According to this mathematical model, we can strengthen other types of slanted-line illusions, such as the Café Wall illusion [11] shown in Fig. 2. In this figure, in each row, white and black rectangles alternate in the horizontal direction, and each row is offset from the adjacent rows by half the width of the rectangles. Although the lines between the rows are straight and horizontal, they appear to be curved and sloped. This illusion can also be explained by the Fermüller model [9].

Fig. 2. Café Wall illusion.

Now, since we know that acute angles will be perceived to be larger than they are, we can expect that this illusion will become stronger if we distort the rectangles into parallelopipeds, since this will generate a series of acute angles. The parallelopiped-based Café Wall pattern is shown in Fig. 3. We can observe that the illusion is stronger, as was predicted by the mathematical model.

These observations, together with the mathematical model, tell us that human visual perception is affected by blurring, even though we feel that we see the figures accurately. We can summarize this observation in the following way.

Observation 1.   The computations in the human brain are imprecise.

Fig. 3. Stronger version of the Café Wall illusion (Sugihara, 2012).

3   Ouchi Illusion and Local Motion Detection

The second example we consider is the Ouchi illusion. The picture shown in Fig. 4 is included in Ouchi's book [16]. This is a still picture, but the central circular area seems to drift at random, independently from the surrounding area. This drift illusion can be explained in the following way.

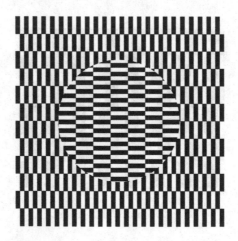

Fig. 4. Ouchi illusion (adopted from [16]).

First, our retina, which is an array of photo sensors, usually has some slight random motion. A sensor generally decreases its sensitivity if it detects the same signal for a relatively long period of time. This is also true of the retina. In order to avoid this decrease in sensitivity, the retina tries to use different sensors to detect a given signal. This is why the retina moves, and hence, when we look at a picture, the image moves on the retina.

Secondly, neurons in the brain, particularly the neurons used in the earlier stages of processing, cover only a small area of the retina. Hence, they process only the local information. This causes ambiguity in the direction of the detected motion, in the following sense.

Suppose that, as shown in Fig. 5, a neuron receives visual data in a small circular area of the retina, and detects displacement of an edge for which both terminal points are outside the circular area. Then, the neuron can tell that the edge moves, but cannot tell in which direction. There are many possibilities for the direction of the motion, as shown by the arrows in Fig. 5. This ambiguity is called the "aperture problem" [13]. In other words, a neuron can only detect motion perpendicular to the edge, no matter which direction the edge actually moves.

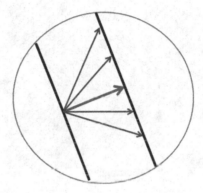

Fig. 5. Aperture problem.

On the basis of these two observations, we can explain why the Ouchi illusion arises. Suppose that the Ouchi pattern moves slightly on the retina. In the central part of the Ouchi pattern, horizontal edges prevail, and, consequently, primarily vertical motion will be detected. In the surrounding part, on the other hand, vertical edges prevail, and, consequently, primarily horizontal motion will be detected. As the result of this, the central and surrounding parts appear to move differently. This is a typical way of explaining the Ouchi illusion [8].

One might think that this illusion would become stronger if the elongated checkerboard patterns were replaced with stripes, because the edge directions would then be more uniform. However, this is not true. The illusion becomes weaker if we replace the central part of the Ouchi pattern with horizontal stripes and the surrounding part with vertical stripes. This can be understood in the following manner. When a neuron becomes excited, it suppresses the excitation of its neighboring neurons. This is called lateral inhibition. If a moving edge is long, the excitation of a neuron covering part of the edge will suppress the excitation of the neurons covering the neighboring parts. In this way, the excitations of neurons cancel each other, and the illusion becomes weaker.

A straight edge will stimulate only those neurons that detect the direction perpendicular to that edge. If the edge direction deviates slightly, such as like a sine curve, it will stimulate more neurons because the edge contains many directions. We can thus expect that the illusion will become stronger if the straight edges are replaced with slightly curved edges.

Based on these observations, we can create patterns that will give a stronger illusion of drift than does the Ouchi pattern. An example of such a pattern is shown in Fig. 6.

Fig. 6. Drift illusion "UFO in the evening glow" (Sugihara, 2013).

From this illusion, we get the next observation of the nature of the human brain.

Observation 2. The basic computations in the human brain are local.

4   Impossible Motion Illusion

Impossible motion is a new type of illusion evoked by a three-dimensional object. Figure 7 shows an example of impossible motion called "magnet-like slopes". Panel (a) shows an object with four slopes. We initially perceive that the four slopes each go down in a different direction from the high center. However, if we place balls on the slopes, they appear to roll uphill toward the high center, defying gravity, as shown in panel (b). Panel (c) shows another view of the same situation; here, we can see that the center is the lowest point, and the balls are just rolling downhill. Thus, the actual motion obeys gravity, but it appears to be an impossible motion that defies physical laws.

This class of visual illusion comes from the fact that a single picture lacks depth information, and so the human brain guesses at the most common solid among infinitely many possibilities that are consistent with what is seen. This

(a)   (b)   (c)

Fig. 7. Impossible motion "Magnet-Like Slopes" (Sugihara, 2009): (a) three-dimensional object; (b) result of apparently impossible motion; (c) another view of the object.

illusion was found as a byproduct of computer vision research [18,19]; refer to [19,21] for the details of designing this class of illusion.

A remarkable aspect of this illusion is that even after we understand the actual shape of the object, as in Fig. 7(c), we incorrectly perceive the shape when we return to the vantage view point shown in Fig. 7(a). This observation may be expressed in the following way.

Observation 3.   Computations in the human brain persistently retain the initial interpretation.

5   Robust Geometric Computations Suggested by the Human Brain

As we have observed, computations in the human brain are imprecise, local, and persistent. However, in spite of these disadvantages, the human brain still can robustly and efficiently process visual geometric data in our daily lives. In this section, we consider how these remarkable characteristics of the human brain can be used in the design of algorithms for computers.

Geometric algorithms are usually designed on the assumption that numerical computations will be done precisely, and hence, in particular, that geometric predicates will always be evaluated correctly. However, this is not true in real computers, and theoretically correct algorithms sometimes fail when they are implemented as software. This failure is common when the input is very close to a degenerate situation.

Let us take the straight skeleton as an example. Let P be a polygon in the plane. Suppose that from each edge of P, two copies of the edge, we will call them the sweep lines, start moving in opposite directions away from the edge and at the same speed, and that they continue to maintain contact with the neighboring sweep lines at the terminal vertices. Hence, the sweep lines change their lengths as they move. The motions of the sweep lines terminate at the points of collision with the other sweep lines. The region swept by a sweep line is assigned to the corresponding edge. In this way, the plane is partitioned into the regions swept for each edge and their boundaries. This partitioning, the boundary structure in particular, is called the *straight skeleton* of the polygon P [2]. Figure 8 shows an example of a polygon (thick lines) and the corresponding straight skeleton (thin lines). The straight skeleton has applications in many fields, such as paper folding [6], solid modeling [27], and pop-up cards [23].

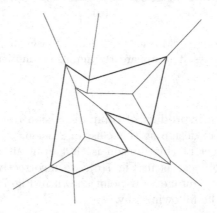

Fig. 8. Polygon and its straight skeleton.

The straight skeleton can be defined for a more general structure, a straight-line graph, and Huber and Held [12] constructed an $O(n^2 \log n)$ algorithm for the generalized straight skeleton. For the straight skeleton of a simple polygon, an $O(n^{3/2} \log n)$ algorithm is known [4]. However, these algorithms are theoretical in the sense that they are designed on the assumption that numerical computation is done precisely. Therefore they are not necessarily valid in actual computers because of numerical errors. In this paper, we apply the topology-oriented principle [22,25] which we have developed for designing robust geometric algorithms, and construct a new algorithm which is robust against numerical errors and which runs in $O(n^2 \log n)$ time, and thus show that the topology-oriented approach is similar to persistent human brain computation.

The straight skeleton can be interpreted as a roof structure in three-dimensional space, in the following manner. Suppose that the polygon P is the shape of the wall seen from above, all parts of the wall have the same height, and we want to construct a roof structure in which all parts have the same angle of declination. The straight skeleton of P tells us how to partition the roof into planar plates so that this is possible. Indeed, to build this roof, the region assigned to an edge of P should be elevated to a roof plate passing through the edge at the top of the wall.

On the basis of this interpretation, we can construct a sweep algorithm for the straight skeleton. Let π be a horizontal plane that is initially placed on top of the wall. We let π sweep upward and, in this way, construct the roof structure inside P step by step from the lower part to the highest point on the roof. Next, we let π sweep downward and thereby construct the remaining part of the roof structure outside of P.

This idea can be summarized in the next algorithm, where we concentrate on the construction of the straight skeleton inside P.

Algorithm 1 (Straight skeleton).

Input: polygon P and angle α between the roof plates and the horizontal plane.
Output: straight skeleton inside P.
Procedure:

1. For each edge e of P, generate the half-plane containing e that is toward the inside of P, forming angle α with respect to the horizontal plane, and put it into storage S. (We will call the elements of S the *roof plates*.)
2. For each vertex v of P, generate the half-line at the intersection of the two roof plates that are associated with the two edges incident to v, and put it into storage E. (We will call the elements of E the *roof edges*.)
3. For each edge e of P, trim the corresponding roof plates in S by the two roof edges emanating from the two terminal vertices of e.
4. Create empty storage locations $\overline{E}$ and $\overline{S}$.
5. Repeat Steps 5.1, 5.2, and 5.3 until E and S are empty.

    5.1 Find a pair (e, s) of a roof edge e and a nonneighboring roof plate s such that they intersect and the point of intersection is the lowest among all such pairs.

    5.2 Move e from E to $\overline{E}$.

    5.3 Increment the roof structure around the point of intersection (details of this procedure will be shown below). If new roof edges are created, add them to E. If the roof plates in S become completely bounded by roof edges, move them from S to $\overline{S}$.

6. Output the roof structure consisting of the roof edges in $\overline{E}$ and the roof plates in $\overline{S}$.   □

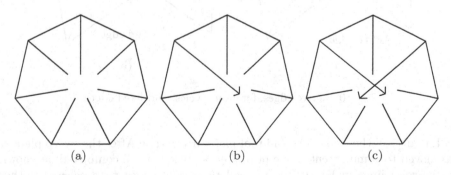

(a)   (b)   (c)

Fig. 9. Inconsistency in the construction of a degenerate straight skeleton: (a) regular polygon and growing skeleton edges; (b) detection of the first vertex; (c) detection of the second vertex, which contradicts the first vertex.

Intuitively, this algorithm constructs the roof structure below the sweep plane π step by step as π moves upward. However, an inconsistency may arise due to numerical errors in the computations. Let P be the regular polygon shown in Fig. 9(a). In theory, all the roof edges emanating from the vertices meet at a common point. However, in the real world, we have numerical errors, and hence it is difficult to identify this common point of intersection. Instead, the algorithm may find many points of intersection. Suppose that Algorithm 1 first finds the point of intersection of a roof edge and a roof plate as shown in Fig. 9(b), and next finds another pair, as shown in Fig. 9(c). However, this is contradictory because the roof edges cross each other, which should not happen in the roof structure. Therefore, Algorithm 1 is not robust against numerical errors.

We can modify this algorithm and make it robust by using what we observed in the previous section about how the brain performs such computation. For this purpose, we first classify the roof edges into three types.

Let us concentrate on the structure below the sweep plane π. As shown in Fig. 10(a), the initial roof edges terminate at the points of intersection with π, and the intersection of the roof plates and π form a cycle represented by broken lines in this figure. We call the roof edges that terminate at the points of intersection with π, the *active edges*, meaning that these edges are still growing. We call the cycle formed by the intersection of the roof plates and π a *forefront cycle*, meaning that it is moving toward the inside of the polygon P. Furthermore, we call the roof plates at the forefront cycles the *active roof plates*, meaning that they are still growing.

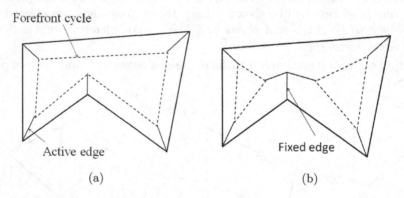

Fig. 10. Active edges, forefront cycles, and fixed edges.

Initially, all the roof edges and roof plates are active. After the sweep plane π has moved to some extent, some of roof edges below π are completed, as shown by the solid lines in Fig. 10(b). We call these roof edges *fixed edges*. In other words, edges in E are active, while edges in $\overline{E}$ are fixed. Similarly, some of the roof plates become completely bounded by roof edges. We call those roof plates *fixed roof plates.* In other words, roof plates in S are active, while those in $\overline{S}$ are fixed.

Once we recognize a forefront cycle, we can identify three types of topological changes in the roof structure that occur during sweeping, as shown in Figs. 11, 12, and 13.

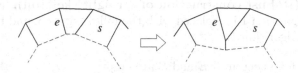

Fig. 11. Event type 1: removal of an edge from a forefront cycle.

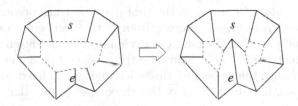

Fig. 12. Event type 2: partition of a forefront cycle.

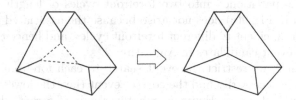

Fig. 13. Event type 3: disappearance of a forefront cycle of length 3.

The first type of event is that the sweep plane π hits a point of intersection between a roof edge e and a roof plate s that is on a roof plate adjacent to s, as shown in Fig. 11. In this case, two active edges become fixed, and the forefront cycle becomes shorter by one.

The second type of event is that the sweep plane π has a point of intersection between a roof edge e and a roof plate s that is not on a plate that is adjacent to the side plate of e, as shown in Fig. 12. In this case, the forefront cycle is partitioned into two, the active e edge becomes fixed, and two new active edges are generated.

The third type of event is that a forefront cycle of length 3 disappears, as shown in Fig. 13. In this case three active edges become fixed, three roof plates become completely bounded by fixed edges, and no new edges are generated.

On the basis of these observations, we can make Algorithm 1 robust by modifying Step 5 in the following way.

Algorithm 2 (Robust construction of straight skeleton).
Step 5 of Algorithm 1 is replaced with the following step, and the remainder is the same as Algorithm 1.

5'. Repeat the following until E and S are empty.
   5.1' Find a pair (e, s) of an active roof edge e and a nonneighboring roof plate s incident to the same forefront cycle, such that their intersection is the lowest.
   5.2' If the event is of type 1, move e and its neighboring edge from E to $\overline{E}$, generate a new active edge, and reconnect the forefront cycle, as shown in Fig. 11. Move from S to $\overline{S}$ the roof plate that has become fixed.
   5.3' If the event is of type 2, move e from E to $\overline{E}$, generate two new active edges, and partition the forefront cycle into two, as shown in Fig. 12.
   5.4' If the event is of type 3, move the three edges (including e) from E to $\overline{E}$, and change the associated three active edges to fixed edges, as shown in Fig. 13. Move from S to $\overline{S}$ the three roof plates that have become fixed. □

Note that this algorithm does not encounter the topological inconsistency shown in Fig. 9. Indeed, at the initial stage (Fig. 9(a)), a forefront cycle of length 7 is generated, and in the first event shown in Fig. 9(b), which is a type-2 event, the forefront cycle is partitioned into two forefront cycles of length 4. Hence, the situation shown in Fig. 9(c) does not arise because the associated roof edge and the roof plate are incident to different forefront cycles, and hence this pair is not included in the event candidates in Algorithm 2.

In Algorithm 2, we restrict the event search to each forefront cycle. In this search, we cannot necessarily find the correct event (i.e., the lowest intersection) because of numerical errors. However, whatever pair of a roof edge and a roof plate is chosen as the next event, no topological inconsistency will arise because the chosen pair will be of type 1, 2, or 3, and thus the topological change of the forefront cycle will be well defined.

In other words, the basic structure of Algorithm 2 allows the topological change of the graph structure of the forefront cycles, and numerical computations are used only to choose the most promising pair of a roof edge and a roof plate. Once this pair is chosen, the algorithm persists in the belief that it gives the lowest intersection and changes the forefront cycles by Steps 5.2', 5.3', or 5.4' of Algorithm 2, depending on the type of the event. Thus we obtain the following theorem.

Theorem 1. Algorithm 2 terminates in finite time and produces a planar graph as output, no matter whether the pair (e, s) chosen in Step 5.1' gives the true lowest intersection or not.

Proof. Once a pair (e, s) is chosen in Step 5.1', the associated event is of type 1, type 2, or type 3, and hence the planar graph consisting of the original polygon edges, active edges, fixed edges, forefront cycle edge, and their terminal vertices is changed by Steps 5.2', 5.3', or 5.4', all of which maintain planarity. Thus, it suffices to show the finiteness of the procedure. Assume that the input polygon P has n edges. Initially, the total number of edges on the forefront cycle is n. This number decreases by one in a type 1 event and by three in a type 3 event. In a type 2 event, the total number of edges on the forefront cycles increases by one, but both of the forefront cycles generated by the partition are smaller by at least two than the forefront cycle before the partition. Hence, Step 5.3' is repeated only a finite number of times, and thus the total number of the forefront cycles eventually vanishes.   □

Theorem 2. Algorithm 2 runs in $O(n^2 \log n)$ time for an n-gon P.

Proof. Steps 1, 2, and 3 are carried out in time $O(n)$, and Step 4 is carried out in time $O(1)$. Step 5' can be performed in the following manner. Initially there are n active edges and n active roof plates. Hence there are $n(n - 2)$ pairs of roof edges and nonneighboring roof plates. We store them in a heap with the y-coordinates of the points of intersection as the keys [1]. We can construct the heap in $O(n^2 \log n)$ time. Deletion of the lowest pair (e, s) from the heap in Step 5.1' requires $O(\log n)$ time. In Steps 5.2' and 5.3', new active edges are generated. As soon as a new active edge e is generated, we compute the point of intersection with each of the nonneighboring roof plates on the same forefront cycle, and add the pair (e, s) to the heap. Adding a pair to the heap requires $O(\log n)$ time. Because there are $O(n)$ roof plates on the same forefront cycle, we can add all the pairs with e to the heap in $O(n \log n)$ time. Hence, Steps 5.2' and 5.3' can be completed in $O(n \log n)$ time. Step 5.4' can be completed in $O(1)$ time. Note that the straight skeleton is a planar graph with n connected regions (corresponding to the n edges of the input polygon) embedded in the plane, and the degree of any vertex is at least three. Hence, the total number of vertices, edges, and connected regions is of $O(n)$. This means that Steps 5.1 to 5.4 are repeated $O(n)$ times. Therefore, Algorithm 2 runs in $O(n^2 \log n)$ time.   □

Figure 14 shows an example of a straight skeleton. The input polygon in this figure has 300 vertices. This polygon was generated by inserting a number of vertices into the edges of a 16-gon and then perturbing their locations with small random numbers. This polygon is not degenerate, and hence the construction of the straight skeleton is not difficult.

Figure 15 shows the output of our algorithm for a regular 30-gon. This is highly degenerate because, if there are no numerical errors, all 30 of the roof edges will meet at the center. In this experiment, the coordinates of the vertices were represented by single-precision floating-point numbers, and the numerical

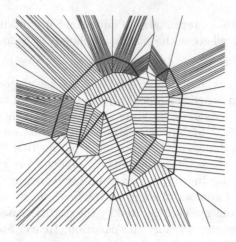

Fig. 14. Straight skeleton constructed by Algorithm 2.

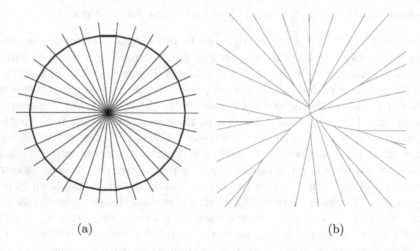

(a)   (b)

Fig. 15. Degenerate straight skeleton: (a) straight skeleton; (b) close-up diagram of the central part.

computations were performed in single-precision floating-point arithmetic. Our algorithm was able to compute the straight skeleton, as shown in Fig. 15(a). However, if we expand the central part by 10^5, we get the diagram in Fig. 15(b), where we can see many vertices instead of a single vertex. This kind of disturbance is not surprising; we note that the algorithm gave a topologically consistent output even though the polygon was highly degenerate.

This algorithm is similar to the computations in the human brain in the sense that both are persistent once a decision has been made, regardless of its accuracy. In this way, both are able to achieve robustness against imprecise numerical computations.

The strategy we followed for designing Algorithm 2 can be considered to be a topology-oriented approach, which we have proposed as a basic principle for designing robust geometric algorithms [22,25]. Indeed, in this approach, the basic part of the algorithm is described in only topological terms, and the numerical data are used only for selecting the most promising branch of the processing. Thus we can say that the idea behind simulating the persistency of the human brain is very similar to the topological approach for robust geometric algorithms.

6 Concluding Remarks

We observed how the human brain processed computations by looking at three visual illusions: the Zöllner illusion, the Ouchi illusion, and the impossible motion illusion, and we then composed a new algorithm for computing straight skeletons. Based on our observations, we pointed out that designing algorithms based on how the human brain computes is very similar to the topology-oriented approach, which we have developed for a long time. Thus, the topology-oriented approach can be used if we want to mimic the processing of the human brain.

Acknowledgment. This research is supported by the Grant-in-Aid for Challenging Exploratory Research No. 24650015 and Scientific Research (B) No. 24360039 of MEXT.

References

1. Aho, A.V., Hopcroft, J.E., Ullmann, J.D.: The Design and Analysis of Computer Algorithms. Addison Wesley, Reading (1974)
2. Aichholzer, O., Aurenhammer, F.: Straight skeletons for general polygonal figures in the plane. In: Cai, J.-Y., Wong, C.K. (eds.) COCOON 1996. LNCS, vol. 1090, pp. 117–126. Springer, Heidelberg (1996)
3. Ballard, D.H., Brown, C.M.: Computer Vision. Prentice Hall, Englewood Cliffs (1982)
4. Cheng, S.-W., Vigneron, A.: Motorcycle graphs and straight skeletons. In: Proceedings of the 13th Annual ACM-SIAM Symposium on Discrete Algorithms, pp. 156–165 (2002)
5. de Berg, M., Cheong, O., van Kreveld, M., Overmars, M.: Computational Geometry: Algorithms and Applications. Springer, Heidelberg (2008)
6. Demaine, E.D., Demaine, M.L., Lubiw, A.: Folding and cutting paper. In: Akiyama, J., Kano, M., Urabe, M. (eds.) JCDCG 1998. LNCS, vol. 1763, pp. 104–118. Springer, Heidelberg (2000)
7. Edelsbrunner, H.: Algorithm in Combinatorial Geometry. Springer, Berlin (1987)
8. Fermüller, C., Pless, R., Aloimonos, Y.: The Ouchi illusion as an artifact of biased flow estimation. Vis. Res. **40**, 77–96 (2000)
9. Fermüller, C., Malm, H.: Uncertainty in visual processes predicts geometric optical illusions. Vis. Res. **44**, 727–749 (2004)
10. Fogaras, D., Sugihara, K.: Topology-oriented construction of line arrangements. IEICE Trans. Fundam. Electron. Commun. Comput. Sci. E85-A, 930–937 (2002)

11. Goto, T., Tanaka, H.: Handbook of the Science of Illusion. University of Tokyo Press, Tokyo (2005)
12. Huber, S., Held, M.: A fast straight-skeleton algorithm based on generalized motor-cycle graphs. Int. J. Comput. Geom. Appl. **22**, 471–498 (2012)
13. Marr, D.: Vision. W. H. Freeman, New York (1982)
14. Ninio, J.: The Science of Illusions. Cornell University Press, Ithaca (2001)
15. Minakawa, T., Sugihara, K.: Topology-oriented construction of three-dimensional convex hulls. Optim. Methods Softw. **10**, 357–371 (1998)
16. Ouchi, H.: Japanese Optical and Geometrical Art. Dover, New York (1977)
17. Preparata, F., Shamos, M.I.: Computational Geometry–An Introduction. Springer-Verlag, New York (1985)
18. Sugihara, K.: Classification of impossible objects. Perception **11**, 65–74 (1982)
19. Sugihara, K.: Machine Interpretation of Line Drawings. The MIT Press, Cambridge (1986)
20. Sugihara, K.: A robust and consistent algorithm for intersecting convex polyhedra. Comput. Graph. Forum 13(Conference Issue), C-45–C-54 (1994). (EUROGRAPH-ICS'94, September 12–16, 1994, Oslo, Norway)
21. Sugihara, K.: A characterization of a class of anomalous solids. Interdisc. Inf. Sci. **11**, 149–156 (2005)
22. Sugihara, K.: Robust geometric computation based on the principle of independence. Nonlinear Theory Appl. IEICE **2**, 32–42 (2011)
23. Sugihara, K.: Design of pop-up cards based on weighted straight skeleton. In: 10th International Symposium on Voronoi Diagrams in Science and Engineering (ISVD 2013), pp. 23–28 (2013)
24. Sugihara, K., Iri, M.: Two design principles of geometric algorithms in finite-precision arithmetic. Appl. Math. Lett. **2**, 203–206 (1989)
25. Sugihara, K., Iri, M.: A robust topology-oriented incremental algorithm for Voronoi diagrams. Int. J. Comput. Geom. Appl. **4**, 179–228 (1994)
26. Sugihara, K., Iri, M., Inagaki, H., Imai, T.: Topology-oriented implementation–an approach to robust geometric algorithms. Algorithmica **27**, 5–20 (2000)
27. Tomoeda, A., Sugihara, K.: Computational creation of a new illusionary solid sign. In: 9th International Symposium on Voronoi Diagrams in Science and Engineering (ISVD 2012), pp. 144–147 (2012)

A Characterization of Link-2 LR-visibility
Polygons with Applications

Xuehou Tan[1,2(✉)], Jing Zhang[1], and Bo Jiang[1]

[1] Dalian Maritime University, Linghai Road 1, Dalian, China
[2] Tokai University, 4-1-1 Kitakaname, Hiratsuka 259-1292, Japan
`tan@wing.ncc.u-tokai.ac.jp`

Abstract. Two points x, y inside a polygon P are said to be mutually *link-2* visible if there exists the third point $z \in P$ such that z is visible from both x and y. The polygon P is *link-2 LR-visible* if there are two points s, t on the boundary of P such that every point on the clockwise boundary of P from s to t is link-2 visible from some point of the other boundary of P from t to s and vice versa. We give a characterization of link-2 LR-visibility polygons by generalizing the known result on LR-visibility polygons. A main idea is to extend the concepts of ray-shootings and components to those under notion of link-2 visibility. Then, we develop an $O(n \log n)$ time algorithm to determine whether a given polygon is link-2 LR-visible.

Using the characterization of link-2 LR-visibility polygons, we further present an $O(n \log n)$ time algorithm for determining whether a polygonal region is searchable by a k-searcher, $k > 2$, improving upon the previous $O(n^2)$ time bound. A polygonal region is searchable by a searcher if the searcher can detect (or see) an *unpredictable* intruder inside the region, no matter how fast the intruder moves. A *k-searcher* holds k flashlights and can see only along the rays of the flashlights emanating from his position. Our result can also be used to simplify the existing solutions of other polygon search problems.

1 Introduction

Let P denote a simple polygon with n vertices. Any two points s and t on P divide the boundary of P into two subchains, which we call L and R, for left and right chains. The LR-visibility question asks whether each point of L is visible from a point of R, and whether each point of R is visible from a point of L. If the answer is *yes*, we say P is LR-visible with respect to s and t. If there exists a pair of points on P such that P is LR-visible with respect to them, we simply say P is LR-visible.

Because of its relation to other problems in polygonal visibility (e.g., the two-guard problem [15] and the polygon search problem [9,12]), the problem of

This work was partially supported by the Grant-in-Aid (MEXT/JSPS KAKENHI 23500024) for Scientific Research from Japan Society for the Promotion of Science and the National Natural Science Foundation of China under grant 61173034.

© Springer International Publishing Switzerland 2014
J. Akiyama et al. (Eds.): JCDCGG 2013, LNCS 8845, pp. 161–172, 2014.
DOI: 10.1007/978-3-319-13287-7_14

characterizing and recognizing LR-visibility polygons has received much attention in computational geometry and robotics [1,3,4,10]. Tseng et al. [15] were
the first to propose an $O(n \log n)$ time algorithm for deciding whether a given
polygon is LR-visible. The time bound was later improved to $O(n)$ by Das et al.
[3]. (One can also report all pairs (s, t) which admit LR-visibility, after determining the LR-visibility of a polygon [3,15].) Very recently, Tan et al. showed
that a polygon P is not LR-visible if and only if it contains k non-redundant
components such that each of them exactly intersects with k' other components,
$0 \leq k' \leq k - 3$, and presented an alternative, simpler algorithm for characterizing LR-visibility polygons [14]. The backward and forward components of a
reflex vertex r are the clockwise and counterclockwise boundary chains from r
to its backward and forward ray-shootings, respectively. A component is *non-
redundant* if it does not contain any other component.

In this paper, we study the link-2 LR-visibility polygons, which are a generalization of LR-visibility polygons. Two points $x, y \in P$ are said to be mutually
link-2 visible if there exists the third point $z \in P$ such that z is visible from
both x and y. A polygon P is *link-2 LR-visible* if there are two points s, t on
the boundary of P such that every point on the clockwise boundary of P from
s to t is link-2 visible from some point of the other boundary of P from t to s
and vice versa.

In the rest of this paper, we first give a characterization of link-2 LR-visibility
polygons. It is obtained by introducing the new concepts of link-2 ray-shootings
and components, and establishing the forbidden patterns for link-2 LR-visibility
polygons, which are a direct generalization of the same structures for LR-
visibility polygons. Next, we develop an $O(n \log n)$ time algorithm to determine
whether a given polygon is link-2 LR-visible. Using the characterization of link-
2 LR-visibility polygons, we further present an $O(n \log n)$ time algorithm for
determining whether a polygonal region is searchable by a k-searcher, $k \geq 2$.
This improves upon the previous $O(n^2)$ time bound [9].

2   Preliminaries

Let P denote a simple polygon with n vertices, i.e., P has neither self-intersections
nor holes. Two points $x, y \in P$ are said to be mutually *visible* if the line segment
connecting them, denoted by $\overline{xy}$, is entirely contained in P. For two regions Q_1,
$Q_2 \subseteq P$, we say that Q_1 is *weakly visible* from Q_2 if every point in Q_1 is visible
from some point in Q_2.

For a vertex x of the polygon P, let $Succ(x)$ denote the vertex immediately
succeeding x clockwise, and $Pred(x)$ the vertex immediately preceding x clockwise. A vertex of P is *reflex* if its interior angle is strictly greater than $180°$;
otherwise, it is *convex*. An important definition for reflex vertices is that of *ray-
shootings*: the backward ray-shooting from a reflex vertex r, denoted by $B(r)$, is
the first point of P hit by a "bullet" shot at r in the direction from $Succ(r)$ to
r, and the forward ray-shooting $F(r)$ is the first point hit by the bullet shot at
r in the direction from $Pred(r)$ to r.

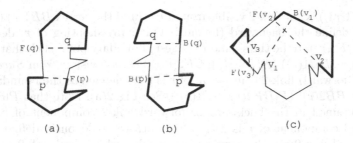

Fig. 1. Forward, backward ray-shootings and components.

Let u, v denote two boundary points of P, and let $P[u,v]$ and $P(u,v)$ denote the closed and open *clockwise* chains of P from u to v, respectively. We define $P[r, B(r)]$ and $P[F(r), r]$ as the *backward component* and *forward component* of the reflex vertex r, respectively. The point r is referred to as the *defining vertex* of its backward or forward component. See Fig. 1. A component is said to be *non-redundant* if it does not contain any other component, no matter which is the forward or backward component. Note that some points of a component of r may not visible from r. The following simple observation can also be made.

Observation 1. *No point outside of the backward (forward) component of a reflex vertex r is visibile from $Succ(r)$ $(Pred(r))$. Morcover, $P(B(r), r)$ $(P(r, F(r))$ is the first boundary chain, immediately after r anti-clockwise (clockwise), which is invisible from $Succ(r)$ $(Pred(r))$.*

The known characterization of LR-visibility polygons is given in terms of the non-redundant components.

Theorem 1 *(See [14]). A polygon P is not LR-visible if and only if it has k non-redundant components such that each of them exactly intersects with k' other components, where $0 \leq k' \leq k - 3$.*

3   Characterizing Link-2 *LR*-visibility Polygons

In this section, we assume that the given polygon P is not LR-visible. We will give a characterization of *link-2 LR*-visibility polygons by generalizing the known result on the LR-visibility polygons (Theorem 1). By introducing the new concepts of link-2 ray-shootings and components, we can generalize the forbidden patterns for LR-visibility polygons into those for link-2 LR-visibility polygons.

Suppose that all non-redundant components in the polygon P have been computed. Given a boundary point p, all boundary points can be ordered in a clockwise scan of P, starting at p. If x is encountered before y, we simply write $x <_p y$. Assume that r is a vertex whose backward component $P[r, B(r)]$ is non-redundant. Let $BB1(r)$ $(BF1(r))$ denote the *largest* (*smallest*) reflex vertex, with respect to r, which is link-2 visible from $Succ(r)$ but $Pred(BB1(r)))$

$(Succ(BF1(r)))$ is not link-2 visible from $Succ(r)$. If the vertex $BB1(r)$ $(BF1(r))$ exists, we define the backward (forward) link-2 ray-shooting of r, denoted by $BB2(r)$ $(BF2(r))$, to be the smallest (largest) boundary point such that no point of $P(BB2(r), BB1(r))$ $(P(BF1(r), BF2(r)))$ is link-2 visible from $Succ(r)$. The backward (forward) link-2 component of r is then defined as the boundary chain $P[BB1(r), BB2(r)]$ $(P[BF2(r), BF1(r)])$. See Fig. 2(a). Note that $P[r, B(r)]$ is always contained in the backward or forward link-2 component of r. Also, if the forward component of r is non-redundant, we analogously define its backward/forward link-2 ray-shooting and the backward/forward link-2 component $P[FB1(r), FB2(r)]/P[FF2(r), FF1(r)]$. See Fig. 2(a). (In the above notion, the first letter "B" (or "F") stands for the original backward (or forward) component of r.) Note that it is possible that only one of the backward and forward link-2 ray-shootings from r is defined. See Fig. 2(b). (Since P is not LR-visible, at least one of the backward and forward link-2 ray-shootings from r is defined.)

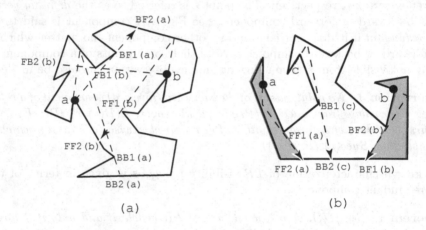

Fig. 2. Link-2 ray-shootings and link-2 components.

We call the link-2 components, which are derived from the non-redundant backward and forward components, the link-2 α-components and β-components, respectively. A link-2 α-component is *non-redundant* if it does not contain any other link-2 α-component, no matter which is the forward or backward α-component. For instance, the α-component $P[BB1(c), BB2(c)]$ in Fig. 2(b) contains the α-component $P[BF2(b), BF1(b)]$ and is thus redundant. (Since P is not LR-visible, both the forward and backward link-2 α-components of a vertex r cannot be non-redundant simultaneously.) We make the analogous definition for the non-redundant β-components.

From the definition of link-2 α-components, the following observation can simply be made. (The observation on the link-2 β-components can be made analogously.)

Observation 2. *Suppose that the backward component of a reflex vertex r is non-redundant, and the backward (forward) link-2 ray-shooting of r is defined. Then, no point outside of the backward (forward) link-2 α-component of r is link-2 visible from $Succ(r)$. Moreover, $P(BB2(r), BB1(r))$ ($P(BF1(r), BF2(r))$) is the first boundary chain, immediately after $BB1(r)$ ($BF1(r)$) anti-clockwise (clockwise), which is not link-2 visible from $Succ(r)$.*

A link-2 component C (it may be an α- or β-component) is said to *intersect* with the other C' if C and C' overlap each other on the boundary of P. We call the endpoint of a link-2 component C the *left* endpoint if it is first encountered, and the other endpoint of C the *right* endpoint if it is second met, in a clockwise scan of the boundary of P, starting at a point that is not contained in C.

The similarity between the components and the link-2 α/β-components (Observations 1 and 2) makes it easy to establish a one-to-one correspondence between the forbidden patterns, for LR-visibility polygons and link-2 LR-visibility polygons.

Lemma 1. *Suppose that the given polygon P is not LR-visible. Then, P is link-2 LR-visible with respect to two boundary points s and t if and only if each of the non-redundant link-2 α-components and β-components contains s or t.*

Proof. The necessity follows from the definition of the link-2 α- and β-components and Observation 2. Assume now that each of the non-redundant link-2 α-components and β-components contains s or t, and thus, each link-2 component contains s or t. A reflex vertex r, say, on the chain L, cannot block $Succ(r)$ nor $Pred(r)$ from being seen from any point of R; otherwise, there exists a link-2 component that contains neither s nor t, a contradiction. The lemma thus follows. □

Lemma 2. *A polygon P is not link-2 LR-visible if it has three disjoint link-2 components.*

Proof. It simply follows from Lemma 1, see Fig. 3(a).   □

Lemma 3. *A polygon P is not link-2 LR-visible if it has k non-redundant link-2 components (they may be the α- or β-components) such that each of them exactly intersects k' other components, where $k \geq 5$ and $k' \leq k - 3$.*

Proof. Figure 3(b) shows an example in which P has five link-2 components; each of them exactly intersects other three components. As in the proof of Lemma 2 of [14], we first transform the k non-redundant link-2 components into k directed chords of a circle R such that the order of chord's endpoints on R is the same as that of the components' endpoints on P. By considering the endpoints of the transformed k chords on R as the vertices of a regular $2k$-polygon, one can easily see that for any two boundary points s and t, there always exists a link-2 component that does not contain s nor t. Thus, P is not link-2 LR-visible. (For details, see the proof of Lemma 2 of [14].)   □

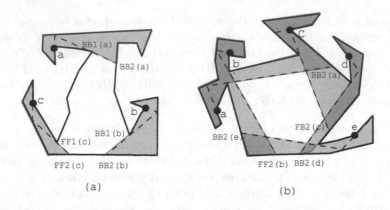

Fig. 3. Two types of the polygons which are not link-2 *LR*-visible.

Theorem 2. *A polygon P is link-2 LR-visible if and only if there do not exist k non-redundant link-2 components (they may be the α- or β-components) such that each of them exactly intersects with k′ other components, where $0 \le k' \le k - 3$.*

Proof. The necessity directly follows from Lemmas 2 and 3. Assume now that there do not exist k non-redundant link-2 components in P such that each of them exactly intersects with k' other components, where $0 \le k' \le k-3$. As in the proof of Theorem 1 of [14], we can classify all non-redundant link-2 components (including both α-components and β-components) into two groups such that the common intersection of the components in either group is not empty. Thus, we can find two boundary points s and t, one per group, such that P is link-2 *LR*-visible with respect to s and t (Lemma 1).   □

4   Recognizing Link-2 *LR*-visibility Polygons

In this section, we present an $O(n \log n)$ time algorithm for determining whether a given polygon P is link-2 *LR*-visible as well as for reporting a pair or all pairs (s, t) which admit link-2 *LR*-visibility. A main procedure is to compute a superset of all non-redundant backward link-2 α-components. A symmetric procedure does the same for the forward link-2 α-components. As described in [3], the non-redundant link-2 α-components can then be extracted from these two sets. Analogously, we compute the non-redundant link-2 β-components. After all non-redundant link-2 components are computed, we can determine whether P is link-2 *LR*-visible (Theorem 2).

By symmetry, we give below only the procedure for computing a superset of non-redundant link-2 backward α-components. We will make use of the shortest path trees, rooted at some polygon vertices. The *shortest path* between two points a and b of P, denoted by $SP(a, b)$, is the Euclidean minimum-distance curve with the endpoints a and b inside P. The path $SP(a, b)$ is always a polygonal chain, whose turning points are the vertices of P. The *shortest path tree* from a vertex v of P is the union of all shortest paths $SP(v, w)$, for each vertex w ($\neq v$) of P. The following two results have been known in the literature.

Lemma 4 *(See [6, 7])*. *The polygon P can be preprocessed in $O(n)$ time so that a shortest path query between two given points can be answered in logarithmic time plus the time required to report the path itself.*

Lemma 5 *(See [2])*. *The polygon P can be preproceessed in $O(n \log n)$ time so that a ray-shooting query can be answered in $O(\log n)$ time.*

4.1 Computing All Non-redundant Backward Link-2 α-Components

We consider below only backward link-2 α-components. Also, we say a backward link-2 α-component is *non-redundant*, if it does not contain any other link-2 backward α-component.

We will first present a method to compute all the backward link-2 α-components, whose defining vertices belong to a non-redundant backward link-2 α-component. The restriction of the defining vertices to a non-redundant link-2 α-component makes it easy to find all the backward link-2 α-components in $O(n \log n)$ time. From Lemma 2, this procedure needs to be performed at most three times.

All Defining Vertices are Limited to a Non-redundant link-2 α-Component. Suppose that the backward link-2 α-component $P[BB1(r), BB2(r)]$ of some vertex r is non-redundant. We describe below a procedure to compute all the backward link-2 α-components, which intersect $P[BB1(r), BB2(r)]$ and whose defining vertices are contained in $P[BB1(r), BB2(r)]$. (Remember that it suffices to compute a *superset* of all non-redundant backward link-2 α-components.) For simplicity, we term these components the backward link-2 α_r-components.

Assume that the backward component of a vertex $p \in P[BB1(r), BB2(r)]$ is non-redundant. If $B(p)$ is visible from $B(r)$, then the backward link-2 ray-shooting of p (e.g., the vertex p' in Fig. 4), if it exists, contains that of r and is thus redundant. Otherwise, the last turning point of the path $SP(B1(r), B(p))$ is the vertex $BB1(p)$, and the link-2 ray-shooting $BB2(p)$ is the boundary point hit by the "bullet" shot at $BB1(p)$ in the direction from $B(p)$ to $BB1(p)$. In this case, $BB2(p) \in P(BB2(r), BB1(r))$ (Fig. 4); otherwise, $P[BB1(p), BB2(p)] \subset P[BB1(r), BB2(r)]$, contradicting the assumption that the link-2 component $P[BB1(r), BB2(r)]$ is non-redundant. In this way, all the backward link-2 α_r-components can be computed.

Consider now the time required to compute the backward link-2 α_r-components. First, all non-redundant backward components in P can be computed $O(n \log n)$ time using the ray-shooting algorithm [2,3] (or even in $O(n)$ time [13]). Next, we compute the shortest paths from $B(r)$ to all the endpoints of the non-redundant backward components, which are contained in $P[BB1(r)$,

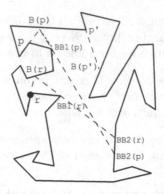

Fig. 4. Computing the link-2 ray-shootings.

$BB2(r)]$ using the linear-time algorithm [5]. The last turing point of a path, say, $SP(B(r), B(p))$, can then be reported in $O(\log n)$ time (Lemma 4), and a link-2 backward ray-shooting can also be computed in $O(\log n)$ time (Lemma 5). Hence, all the backward link-2 α_r-components can be calculated in $O(n \log n)$ time.

Lemma 6. *For a non-redundant backward link-2 component of r, we can compute in $O(n \log n)$ time all the backward link-2 α_r-components, whose defining vertices belong to the backward link-2 α-component of r.*

Calling the Procedure of Computing the Link-2 α_r-Components Three Times. It follows from Lemma 2 that the procedure given in the previous section needs to be performed for at most three non-redundant backward link-2 α-components; either a superset of the backward link-2 α-components is eventually computed, or three (pairwise) disjoint components are found. In the latter case, P is not link-2 LR-visible.

Let us now describe how to find three non-redundant backward link-2 α-components, to which the above procedure applies. Assume first that the backward component of a reflex vertex r is non-redundant. The backward link-2 component of r can then be found from the visibility polygon of the line segment $\overline{rB(r)}$. The visibility polygon of $\overline{rB(r)}$ in P can be computed in linear time [5]. Assume also that the backward link-2 α-component $P[BB1(r), BB2(r)]$ is found. See Fig. 5. Next, find the first reflex vertex v, by a counterclockwise scan from $BB2(r)$ to $BB1(r)$, such that $P[BB1(v), BB2(v)]$ is contained in $P[BB1(r), BB2(r)]$ (Fig. 5). From Lemmas 4 and 5, it can simply be done in $O(n \log n)$ time. If no vertex v exists, the backward link-2 α-component of r is non-redundant, with respect to all backward link-2 α-components (Fig. 5(c)). Otherwise, since v is the first vertex, in the counterclockwise scan from $BB2(r)$ to $BB1(r)$, such that $P[BB1(v), BB2(v)]$ is contained in $P[BB1(r), BB2(r)]$, the backward link-2 α-component of v is non-redundant. See Figs. 5(a)–(b). (The backward link-2 α-component of r may be redundant, see Fig. 5(a).)

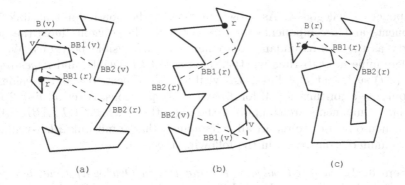

Fig. 5. Illustration for computing the non-redundant backward link-2 α-components.

Without loss of generality, assume that the backward link-2 α-component of r is non-redundant. Then, all backward link-2 α_r-components are computed. Next, we compute the first vertex u counterclockwise (if it exists) such that its backward link-2 α-component is non-redundant and disjoint from that of r. As described above, the backward link-2 α-component of u (if it exists) can be found in $O(n \log n)$ time, by a counterclockwise scan from $BB2(r)$ to $BB1(r)$. If the vertex u does not exist, we then compute the backward link-2 α-components, whose defining vertices belong to $P[BB2(r), BB1(r)]$. Clearly, all found components give a superset of the non-redundant backward link-2 α-components. In the case that the link-2 backward α-component of u is found, we compute the backward link-2 α_u-components. Finally, we further find the vertex z, immediately succeeding u, such that the backward link-2 α-component of z is non-redundant. If z happens to be identical to r, then the backward link-2 α_r-components and α_u-components give a superset of the non-redundant backward link-2 α-components, and we are done. If the backward link-2 α-component of z is disjoint from those of both u and v, we report that P is not link-2 LR-visible (Lemma 2). Otherwise, we also compute the backward link-2 α_z-components. The union of all found components clearly gives a superset of the non-redundant backward link-2 α-components. (See also [14] for a detailed description of the same purpose procedure for LR-visibility polygons.)

Therefore, we can conclude the following result.

Lemma 7. *For a simple polygon P with n vertices, one can in $O(n \log n)$ time compute a superset of non-redundant backward link-2 α-components or report that P is not link-2 LR-visible.*

4.2   The Algorithm

We first compute two supersets of the backward and forward non-redundant link-2 α-components, and then extract the exact set of non-redundant link-2 α-components from them [3]. (If it is ever reported that P is not link-2 LR-visible, then we are done.) Also, we compute all the non-redundant link-2

β-components analogously. As shown in Sect. 3, the non-redundant link-2 α-components and β-components for link-2 LR-visible polygons have the same property as the non-redundant components for LR-visible polygons. We can then determine in $O(n)$ time whether P is link-2 LR-visible, and if *yes* report a pair (s,t) such that P is link-2 LR-visible with respect to s and t, using the same purpose algorithm of [14] for LR-visibility polygons. After all link-2 non-redundant components are computed, the algorithm $ALL\_2\text{-}CUTTABL\_PAIRS$ of [15] can also be used to report all pairs s and t that admit link-2 LR-visibility.

We summarize our result in the following theorem.

Theorem 3. *For a given polygon P, one can in $O(n \log n)$ time determine whether P is link-2 LR-visible, and if so, report a pair or all pairs (s,t) which admit link-2 LR-visibility.*

5   Applications

In this section, we discuss on the application of our characterization of link-2 LR-visibility polygons to the *polygon search problem* and other visibility problems as well.

A polygonal region P is said to be *searchable* by a searcher if the searcher can detect (or see) an *unpredictable* intruder inside the region, no matter how fast the intruder moves. A *k-searcher* holds k flashlights and can see only along the rays of the flashlights emanating from his position. It has been known that searchability of an ∞-searcher is the same as that of a 2-searcher [9]. This result is obtained by characterizing the polygons which are searchable by a k-searcher, $k \geq 2$. An $O(n^2)$ time solution to the polygon search problem for a 2-searcher was then claimed in [8]. Note that P is not searchable by a 1-searcher (a 2-searcher) if P is not LR-visible (link-2 LR-visible) [12].

In the following, we present an $O(n \log n)$ time algorithm for determining whether a polygonal region is searchable by a 2-searcher. Our results improves upon the previous $O(n^2)$ time bound [9]. To this end, we first recall the characterization of the 2-searchable polygons in terms of the non-redundant link-2 components (which may be the α- or β-components.)

Lemma 8. *(See [9].) A polygon P is searchable by a 2-searcher, if and only if none of the following conditions holds:*

A1 *There are three link-2 components in P such that one forward component intersects the other backward component and the third is disjoint from both the intersecting components (see Fig. 6(a)).*

A2 *The polygon P is not link-2 LR-visible.*

A3 *For any boundary point p, there are a backward link-2 component C and a forward link-2 component C' such that p is not contained in both C and C' (see Fig. 6(b)).*

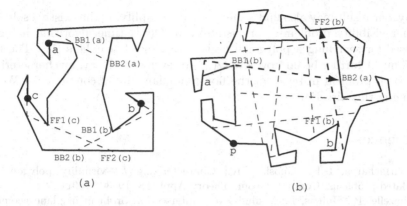

Fig. 6. Two types of the link-2 LR-visible polygons which are not 2-searchable.

Remark. The conditions **N1**, **N2** and **N3** given in [9] for characterizing the k-searchable polygons make use of the concept of *essential cuts*, which are essentially the same as the non-redundant components. The condition **N1** consists of **A1** and one more situation of three disjoint link-2 components (Lemma 2), and **N2** describes the same situation as stated in Lemma 3. The condition **N3** is essentially the same as **A3**. Thus, the forbidden situations described by the conditions **N1**, **N2** and **N3** in [9] are the same as those by **A1**, **A2** and **A3**. Note also that the polygons shown in Fig. 6 are link-2 LR-visible.

First, we can check in $O(n \log n)$ time whether P is link-2 LR-visible (Theorem 3). If *not*, then P is not 2-searchable. Assume now that P is link-2 LR-visible. It is easy to see that all the components satisfying **A3** are essentially non-redundant. Also, if there are three components satisfying **A1**, then they are non-redundant; otherwise, there are three pairwise disjoint link-2 components, contradicting the assumption that P is link-2 LR-visible.

Suppose that all non-redundant backward (forward) link-2 components (they may be the α- or β-components) have been computed, and their endpoints are ordered by a clockwise wise of P, starting at an arbitrary boundary point. For a boundary point p, we can in $O(1)$ time find a non-redundant backward (forward) link-2 component C (C'), whose left (right) endpoint is closest to p clockwise (counterclockwise), starting at p. By scanning on the boundary of P once, we can then determine in $O(n)$ time whether the condition **A3** is satisfied. Analogously, for a non-redundant component C (e.g., the link-2 component of a in Fig. 6(a)), we can also in $O(1)$ time find a non-redundant forward (backward) link-2 component D (D'), which is closest to the left (right) endpoint of C in clockwise (counterclockwise) direction. Such a triple of the components, if it exists, can also be found in $O(n)$ time. Hence, whether the condition **A1** is satisfied in P can be verified in $O(n)$ time. Putting together all the above results, we obtain the following result.

Theorem 4. *For a given polygon P, one can determine in $O(n \log n)$ time whether P is searchable by a k-searcher, where $k (\geq 2)$ is a fixed, positive integer.*

Finally, our result may also find applications in simplifying the existing solutions of other visibility problems. For instance, an $O(n^4)$ time algorithm has been proposed for searching a polygonal region with two 1-searchers [11]. The time bound might further be improved to $O(n^3)$ or even $O(n^2)$, say, by characterizing this search problem in terms of the non-redundant link-2 components. We are working in this direction.

References

1. Bhattacharya, B.K., Ghosh, S.K.: Characterizing LR-visibility polygons and related problems. Comput. Geom. Theory Appl. **18**, 19–36 (2001)
2. Chazelle, B., Guibas, L.: Visibility and intersection problem in plane geometry. Discrete Comput. Geom. **4**, 551–581 (1989)
3. Das, G., Heffernan, P.J., Narasimhan, G.: LR-visibility in polygons. Comput. Geom. Theory Appl. **7**(1), 37–57 (1997)
4. Ghosh, S.K.: Visibility Problems in the Plane. Cambridge University Press, Cambridge (2007)
5. Guibas, L., Hershberger, J., Leven, D., Sharir, M., Tarjan, R.: Linear time algorithms for visibility and shortest path problems inside triangulated simple polygons. Algorithmica **2**, 209–233 (1987)
6. Guibas, L.J., Hershberger, J.: Optimal shortest path queries in a simple polygon. J. Comput. Syst. Sci. **39**, 126–152 (1989)
7. Hershberger, J.: A new data structure for shortest path queries in a simple polygon. Inform. Process. Lett. **38**, 231–235 (1991)
8. Lee, J.H., Park, S.M., Chwa, K.Y.: Simple algorithms for searching a polygonal region with flashlights. Inform. Process. Lett. **81**, 265–270 (2002)
9. Park, S.-M., Lee, J.-H., Chwa, K.-Y.: Visibility-based pursuit-evasion in a polygonal region by a searcher. In: Orejas, F., Spirakis, P.G., van Leeuwen, J. (eds.) ICALP 2001. LNCS, vol. 2076, pp. 456–468. Springer, Heidelberg (2001)
10. O'Rourke, J.: Computational Geometry in C. Cambridge University Press, Cambridge (1993)
11. Simov, B.H., Slutzki, G., LaValle, S.M.: Clearing a polygon with two 1-searchers. Int. J. Comput. Geom. Appl. **19**, 59–92 (2009)
12. Suzuki, I., Yamashita, M.: Searching for mobile intruders in a polygonal region. SIAM J. Comp. **21**(5), 863–888 (1992)
13. Tan, X.: A linear-time 2-approximation algorithm for the watchman route problem for simple polygons. Theor. Compt. Sci. **384**(1), 92–103 (2007)
14. Tan, X., Jiang, B., Zhang, J.: Characterizing and recognizing LR-visibility polygons. Discrete Appl. Math. **165**, 303–311 (2014)
15. Tseng, L.H., Heffernan, P.J., Lee, D.T.: Two-guard walkability of simple polygons. Int. J. Comput. Geom. Appl. **8**(1), 85–116 (1998)

Imaginary Hypercubes

Hideki Tsuiki$^{(\boxtimes)}$ and Yasuyuki Tsukamoto

Graduate School of Human and Environmental Studies,
Kyoto University, Kyoto, Japan
{tsuiki,tsukamoto}@i.h.kyoto-u.ac.jp

Abstract. Imaginary cubes are three-dimensional objects that have square projections in three orthogonal ways, just like a cube has. In this paper, we introduce higher-dimensional extensions of imaginary cubes and study their properties.

1 Introduction

Imaginary cubes are three-dimensional objects that have square projections in three orthogonal ways, just like a cube has [1]. A regular tetrahedron and a cuboctahedron are examples of imaginary cubes (Fig. 1(a,b)). There are two imaginary cubes with remarkable geometric properties: a hexagonal bipyramid imaginary cube (Fig. 1(c); we simply call it an H) and a triangular antiprismoid imaginary cube (Fig. 1(d); we call it a T). Figure 2 shows how they can be considered as imaginary cubes. The first author of this paper has studied imaginary cubes, in particular minimal convex imaginary cubes and fractal imaginary cubes. He has also designed sculptures and puzzles based on them [1–4].

In this paper, we study higher-dimensional extensions of imaginary cubes. In particular, we study n-dimensional counterparts of regular tetrahedron, H, and T for each $n \geq 2$, which we call S^n, H^n, and T^n, respectively. We also study fractal imaginary cubes that correspond to these three series of polytopes.

In Sect. 2, we review properties of imaginary cubes based on [1]. Then, we study higher-dimensional extensions of them in Sect. 3, and fractal imaginary hypercubes in Sect. 4.

Objects and Polytopes

Here, we only study imaginary cubes that are compact subsets of $\mathbb{R}^n$. Therefore, an *object* means a non-empty compact subset of $\mathbb{R}^n$ in this paper. We say that two objects are *similar* if one can be transformed to the other by scaling and isometry. We call this equivalence class a *shape*. Each shape S is also regarded as a name of an object, and we say that an object A is an S if A belongs to the class S. We use roman font to denote a shape, but italic font is used for objects.

A *polytope* is a convex hull of a finite set of points in $\mathbb{R}^n$. We denote by conv(A) the convex hull of an object A, and by vert(P) the set of vertices of a polytope P. A *facet* of an n-dimensional polytope P is an $(n-1)$-dimensional

© Springer International Publishing Switzerland 2014
J. Akiyama et al. (Eds.): JCDCGG 2013, LNCS 8845, pp. 173–184, 2014.
DOI: 10.1007/978-3-319-13287-7_15

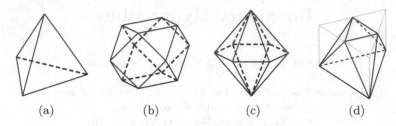

Fig. 1. Examples of imaginary cubes: (a) regular tetrahedron, (b) cuboctahedron, (c) H: hexagonal bipyramid with 12 isosceles triangle faces with a height 3/2 of the base, (d) T: triangular antiprismoid obtained by truncating the three vertices of a base of a regular triangular prism whose height is $\sqrt{6}/4$ of an edge.

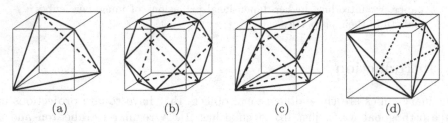

Fig. 2. Imaginary cubes in Fig. 1 placed in cubes.

face of P. We simply call an n-dimensional hypercube an *n-cube*. We refer the reader to [5] for background material on polytopes.

For any two objects A and B, and for any scalar $c \in \mathbb{R}$, we set their Minkowski sum $A + B = \{a + b \in \mathbb{R}^n \mid a \in A, b \in B\}$, and scaling $cA = \{ca \mid a \in A\}$. In this paper, $\mathbf{1}$ is the vector $(1, \ldots, 1) \in \mathbb{R}^n$, and "·" is the dot product on $\mathbb{R}^n$.

2   Imaginary Cubes

Imaginary cubes are three-dimensional objects with square projections in three orthogonal ways. Note that a regular octahedron also has square projections in three orthogonal ways, but its square projections are arranged differently. We exclude such a case by defining an imaginary cube more precisely as follows.

Definition 1. Let C be a 3-cube, and A be an object.

1. A is an *imaginary cube of C* if A has the same three square projections as C has.
2. A is an *imaginary cube* if it is an imaginary cube of a cube.
3. A is a *minimal convex imaginary cube (MCI for short) of C* if A is minimal among convex imaginary cubes of C.
4. A is an *MCI* if it is an MCI of a cube.

It is clear that a convex object A is an imaginary cube of C if and only if each edge of C contains at least one point of A. Therefore, an MCI of C is a convex hull of some points of the edges of C, and thus it is a polytope.

Let A be an MCI of a cube C. The vertices of A are divided into two categories: *v-vertices*, which are also vertices of C, and *e-vertices*, which are not vertices of C. We denote by $V(A)$ the set of v-vertices of A.

Definition 2. A *0/0.5/1 MCI of* C is an MCI with its e-vertices at middle points of the edges of C.

Each object in Fig. 1 is a 0/0.5/1 MCI. Note that a regular tetrahedron has only v-vertices and a cuboctahedron has only e-vertices.

For a polytope A, a subset of vert(A) is called a star if it is composed of a vertex and all of its adjacent vertices.

Theorem 3 (Theorem 3 and Corollary 4 of [1]). *There is one-to-one correspondence between 0/0.5/1 MCIs of* C *and subsets of* vert(C) *that do not contain any star as their subset. There are 15 0/0.5/1 MCI shapes.*

Proof. For an MCI A of C, $V(A)$ does not contain any star because of its minimality. On the other hand, from a subset $S \subset$ vert(C) without a star, we obtain an MCI by selecting its e-vertices on middle points of the edges of C both of whose endpoints are not in S.

There are 15 equivalence classes of subsets of vert(C) without a star. Here, two subsets of vert(C) are equivalent if one is transformed to the other by an isometry which fixes C. We can easily check that every pair of them induces non-similar 0/0.5/1 MCIs. Therefore, there are 15 0/0.5/1 MCI shapes.   □

We say that two MCIs A and A' of C are *v-equivalent* if $V(A)$ can be transformed to $V(A')$ by an isometry which fixes C. There is a representative 0/0.5/1 MCI in each v-equivalence class. The list of all 0/0.5/1 MCIs is given in [1].

We define a *double imaginary cube* as an imaginary cube of two different cubes. As Fig. 3 shows, an H (Fig. 1(c)) is the intersection of two cubes and is a double imaginary cube. It is shown that all the convex double imaginary cubes are intersections of two cubes of the same size which share a diagonal and thus they are MCIs v-equivalent to H [1, Proposition 5].

We call an n-dimensional polytope with $2n$ vertices a *weak cross-polytope* if its vertices are on the positive and the negative sides of a set of axes of coordinates,

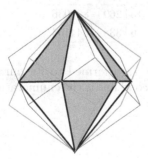

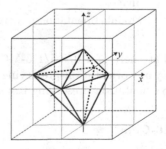

Fig. 3. H as the intersection of two cubes.

Fig. 4. T as a weak polytope.

and call it a *cross-polytope* if the distances from the origin to the vertices are the same. As Fig. 4 shows, a T (Fig. 1(d)) is a three-dimensional weak cross-polytope as well as an imaginary cube.

Hs and Ts form a tiling of three-dimensional Euclidean space and this tiling is closely related to the properties that H is a double imaginary cube shape and T is a weak cross-polytope imaginary cube shape. We explain this tiling in Sect. 3.4 together with another tiling by imaginary hypercubes.

3   Imaginary Hypercubes

3.1   Minimal Convex Imaginary n-cubes

We extend the theory of imaginary cubes to higher-dimensional cases. For $n \geq 2$, we say that an object A is an *imaginary n-cube of an n-cube* C if A has $(n-1)$-cube projections in n orthogonal directions, just like C has. An *imaginary n-cube* is defined as an imaginary n-cube of some n-cube.

We define the following as we did in the three-dimensional case: a minimal convex imaginary n-cube (n-MCI for short) (of C), a 0/0.5/1 MCI (of C), v-vertices and e-vertices of an n-MCI (of C), and v-equivalence on n-MCIs of C. We omit the dimension when it is obvious from the context, and an n-MCI is called an MCI, for example.

Theorem 4. *For an n-cube C with $n \geq 2$, there is a one-to-one correspondence between 0/0.5/1 MCIs of C and subsets of* vert(C) *without a star.*

The proof of this theorem is the same as that of Theorem 1 and is omitted.

Also in higher-dimensional cases, some objects are 0/0.5/1 MCIs of two different n-cubes. However, the set of v-vertices of such an object does not depend on the choice of the cube as we will show in Theorem 9. Therefore, we only have to enumerate equivalence classes of subsets of vert(C) without a star in order to enumerate 0/0.5/1 MCI shapes. We calculated these numbers for the case $n \leq 5$ with a computer program.

n	2	3	4	5
Shapes	4	15	269	829036
Modulo orientation-preserving isometries	4	16	338	1544164

Note that there is a 0/0.5/1 3-MCI that cannot be transformed to its mirror image by any orientation-preserving isometry. The second line is the enumeration modulo orientation-preserving isometries.

3.2   16-Cells

A 16-cell is a four-dimensional cross-polytope. It is a four-dimensional regular polytope with 16 regular tetrahedron facets. See [8], for example, about properties of regular polytopes.

Let A_1 be a 16-cell given by $\mathrm{vert}(A_1) = V_1 = \{(\pm1, 0, 0, 0), (0, \pm1, 0, 0),$ $(0, 0, \pm1, 0), (0, 0, 0, \pm1)\}$, and let $C_1 = \mathrm{conv}(\{-1/2, 1/2\}^4)$ be a 4-cube. Let V_2 and V_3 be the subsets of $\mathrm{vert}(C_1)$ with even and odd numbers of $1/2$-coordinates, respectively. One can see that $C_2 = \mathrm{conv}(V_1 \cup V_3)$ and $C_3 = \mathrm{conv}(V_1 \cup V_2)$ are also 4-cubes. Since V_1 does not contain any star of C_2 (resp. C_3), and every edge of C_2 (resp. C_3) contains a point in V_1, we can find that A_1 is an imaginary cube of C_2 (resp. C_3) that has no e-vertices. Thus, A_1 is a double imaginary 4-cube. Note that $A_2 = \mathrm{conv}(V_2)$ and $A_3 = \mathrm{conv}(V_3)$ are also 16-cells.

As we described in Sect. 2, T is a weak cross-polytope imaginary cube shape and H is a double imaginary cube shape in the three-dimensional case. We show that 16-cell is the only weak cross-polytope imaginary cube shape as well as the only double imaginary cube shape in four and higher-dimensional cases. First, we study weak cross-polytope imaginary cubes.

Lemma 5. *A convex imaginary n-cube polytope has at least 2^{n-1} vertices.*

Proof. An n-cube has $n2^{n-1}$ edges and a convex imaginary n-cube polytope contains a vertex on each of these edges. Since a vertex is on at most n edges of the cube, we have the result. □

Proposition 6. *For $n \geq 3$, T and 16-cell are the only weak cross-polytope imaginary hypercube shapes.*

Proof. Since an n-dimensional weak cross-polytope has $2n$ vertices, n must satisfy $2n \geq 2^{n-1}$ by Lemma 5, and hence $n \leq 4$.

For $n = 4$, any weak cross-polytope has eight vertices. If a weak cross-polytope A is an imaginary cube polytope of a 4-cube C, then A is a MCI with no e-vertices, and each edge of C contains one vertex of A from the proof of Lemma 5. Thus A is a 16-cell.

For $n = 3$, assume that a weak cross-polytope A is an imaginary cube of a 3-cube C. Note that A may not be an MCI of C. We set $V(A) := \mathrm{vert}(C) \cap A$. Since a 3-cube has 12 edges and A has 6 vertices, we have $3\#V(A) + (6 - \#V(A)) \geq 12$, and get $\#V(A) \geq 3$.

If $\#V(A) = 3$, A has three e-vertices and they must be on the edges of C both of whose endpoints are not in $V(A)$. Thus, A is an MCI of C, and we can find that T is the only such polytope.

If $\#V(A) \geq 4$, there is a pair $\{v_1, v_2\} \subset V(A)$ such that $\mathrm{vert}(A) \setminus \{v_1, v_2\}$ is on a plane that is orthogonal to the line segment $[v_1, v_2]$. Suppose that $C = \mathrm{conv}\{0, 1\}^3$. Since $[v_1, v_2]$ contains an interior point of C, we can put $v_1 = (0, 0, 0)$ and $v_2 = (1, 1, 1)$ without loss of generality. Suppose that the other four vertices of A are on a plane defined as $\{(x, y, z) \mid x + y + z = a\}$ $(a \in \mathbb{R})$. Since A has four or more v-vertices, we get $a = 1, 2$. If $a = 1$, $\mathrm{vert}(A)$ must contain $(1, 0, 0), (0, 1, 0)$ and $(0, 0, 1)$. However, no line passes through two of them and the origin $(1/3, 1/3, 1/3)$ at the same time. Therefore, we have no weak cross-polytope in this case. The case $a = 2$ is similar to the case $a = 1$. □

Next, we study double imaginary n-cubes. A convex object can be an imaginary 2-cube of two or more squares. For example, a square is an imaginary cube of infinitely many squares. In the three-dimensional case, there are many convex double imaginary cubes, and H is the only 0/0.5/1 MCI among them as we mentioned in Sect. 2. For $n \geq 4$, we show that 16-cell is the only convex double imaginary n-cube shape. We prepare two lemmas, whose proofs are omitted.

Lemma 7. *For $n \geq 3$, the dimension of the affine hull of an imaginary n-cube is n.*

Note that this lemma does not hold for $n = 2$ because a line segment is an imaginary 2-cube.

For an n-dimensional hyperplane G, we denote by $r(G)$ he distance of G from the origin.

Lemma 8. *Let $C = \mathrm{conv}(\{-1, 1\}^n)$ and G be an n-dimensional hyperplane.*

(1) If $n > 4$ and one of the open half spaces defined by G contains only one vertex of C, then $r(G) > 1$.

(2) If $n = 4$ and one of the open half spaces defined by G contains only one vertex v of C, then $r(G) \geq 1$. If $r(G) = 1$, in addition, then the four adjacent vertices of v are on G.

Theorem 9. *16-cell is the only convex double imaginary 4-cube shape. For $n > 4$, there is no double imaginary n-cube.*

Proof. Let $n \geq 4$. Suppose that B is a double imaginary cube of two n-cubes C_1 and C_2. One can see that $A = C_1 \cap C_2$ is a convex double imaginary cube because we have $B \subset A$. We consider the double imaginary cube A.

We can assume without loss of generality that $C_1 = \mathrm{conv}(\{-1, 1\}^n)$ and that the edge length of C_2 is less than or equal to the edge length of C_1, that is, 2. Let P be a facet of C_2 and G be the hyperplane containing P. All the edges of P must intersect with C_1 because A is an imaginary cube of C_2. Hence $P \cap C_1$ is an imaginary cube of an $(n-1)$-cube P. Since $n \geq 4$, the dimension of its affine hull is $n - 1$ by Lemma 7. On the other hand, it is immediate to show that each facet of C_1 is not on G. Therefore, there exists a vertex v of C_1 in the open half-space defined as the opposite side of C_2 with respect to G. Such a vertex of C_1 is unique because every edge of C_1 must intersect with C_2. Therefore, if $n > 4$, then $r(G) > 1$ by Lemma 8. Since it also holds for the facet P' which is parallel to P, the edge length of C_2 is greater than 2, contradicting the assumption. Therefore, we have $n = 4$. By Lemma 8, the two 4-cubes have the same size and P contains all the four adjacent vertices of v. Therefore, $P \cap C_1$ is a regular tetrahedron. Since C_1 and C_2 have the same size, it holds for all the facets of C_1 and C_2. Therefore, A is a 16-cell. Since a 16-cell is a minimal convex imaginary 4-cube, it is the only convex double imaginary 4-cube.   □

3.3 Higher Dimensional Extensions of H and T

In this subsection, we make n-dimensional extensions of the four 0/0.5/1 MCIs in Fig. 1 in each $n \geq 2$. We regard the 0/0.5/1 n-MCI which has no v-vertices as an imaginary n-cube corresponding to a cuboctahedron.

As an n-dimensional counterpart of a regular tetrahedron, we define S^n and S'^n as follows:

$$V(S^n) = \{x \in \{0,1\}^n \mid x \cdot 1 \equiv 0 \pmod{2}\},$$
$$V(S'^n) = \{x \in \{0,1\}^n \mid x \cdot 1 \equiv 1 \pmod{2}\}.$$

Let $x, y \in \{0,1\}^n$ be two vertices of $C = \text{conv}(\{0,1\}^n)$. If x and y are the two endpoints of an edge of C, we get $x \cdot 1 = y \cdot 1 \pm 1$. Therefore, every edge of C contains points of both S^n and S'^n. Therefore, S^n and S'^n are imaginary cubes of C that have no e-vertices. Moreover, since both $V(S)$ and $V(S')$ contain no star of C, S and S' are MCIs of C. Note that S^n and S'^n have the same shape which is denoted by S^n. The shape S^4 is 16-cell.

Concerning H and T, we define three 0/0.5/1 MCIs of an n-cube $C = \text{conv}(\{-1,1\}^n)$ as follows:

$$V(H^n) = \{x \in \{-1,1\}^n \mid x \cdot 1 \equiv 0 \pmod{3}\},$$
$$V(T^n) = \{x \in \{-1,1\}^n \mid x \cdot 1 \equiv -1 \pmod{3}\}, \qquad (1)$$
$$V(T'^n) = \{x \in \{-1,1\}^n \mid x \cdot 1 \equiv 1 \pmod{3}\}.$$

By a similar argument, one can see that they define 0/0.5/1 MCIs. Note that T^n and T'^n are similar because we have $T^n = -T'^n$. We denote by H^n and T^n the shapes of H^n and T^n, respectively.

These sets of vertices satisfy the following equations. We have

$$V(H^{n+1}) = V(T'^n) \times \{-1\} \cup V(T^n) \times \{1\},$$
$$V(T^{n+1}) = V(H^n) \times \{-1\} \cup V(T'^n) \times \{1\}, \qquad (2)$$
$$V(T'^{n+1}) = V(T^n) \times \{-1\} \cup V(H^n) \times \{1\}.$$

One can see from (1) that each of H^n, T^n and T'^n is mapped to itself by a permutation of the n coordinates. Therefore, one can derive from Eq. (2) that for $n \geq 4$, H^n has $2n$ copies of T^{n-1} facets. The other facets are $(n-1)$-simplexes because each vertex figure of an n-cube is a simplex. On the other hand, T^n has n copies of H^{n-1} facets, n copies of T^{n-1} facets and some $(n-1)$-simplex facets for $n \geq 4$. In the case $n = 3$, the six 2-simplex facets of H^3 coincide with T^2 and the three H^2 facets of T^3 degenerate to line segments. Thus, H^3 has twelve T^2 faces and T^3 has eight faces.

One can see that the set of e-vertices of H^n, T^n and T'^n are the sets

$$\{x \in \{-1,0,1\}^n \mid x \cdot 1 \equiv 0 \pmod{3}, \ x \cdot x = n - 1\},$$
$$\{x \in \{-1,0,1\}^n \mid x \cdot 1 \equiv -1 \pmod{3}, \ x \cdot x = n - 1\}, \text{ and} \qquad (3)$$
$$\{x \in \{-1,0,1\}^n \mid x \cdot 1 \equiv 1 \pmod{3}, \ x \cdot x = n - 1\},$$

respectively.

3.4   Tilings by Imaginary Cubes

As we mentioned above, Hs and Ts form a tiling of three-dimensional Euclidean space, and 16-cells form a tiling of four-dimensional Euclidean space. We explain these tilings from the viewpoints of weak cross-polytope imaginary cubes and double imaginary cubes.

We set positive integers $n \geq 3$ and $k \geq 2$. Consider a subset Z of the n-dimensional cubic lattice

$$Z = \{x \in \mathbb{Z}^n \mid x \cdot 1 \equiv 0 \pmod{k}\}.$$

We call a cube $\mathrm{conv}(\{0,1\}^n) + \{v\}$ $(v \in \mathbb{Z}^n)$ a *lattice-cube*. In each lattice-cube C, take an MCI of C whose set of v-vertices is $Z \cap \mathrm{vert}(C)$. Such an MCI is a translation of one of M_r for $0 \leq r < k$ defined as

$$V(M_r) = \{x \in \{0,1\}^n \mid x \cdot 1 \equiv r \pmod{k}\}.$$

Note that every pair of these MCIs which are placed in adjacent n-cubes share the faces on their intersection. After placing such MCIs, there remain holes around lattice points

$$\{x \in \mathbb{Z}^n \mid x \cdot 1 \not\equiv 0 \pmod{k}\}.$$

These holes are weak cross-polytopes because all of the vertices are on the lattice edges. Therefore, for every n and k, we have a tiling of n-dimensional space by translations of M_r for $0 \leq r < k$ and n-dimensional weak cross-polytopes of several shapes. In the case $n = 3$ and $k = 2$, this tiling is the three-dimensional tiling by regular tetrahedra and regular octahedra. In the case $n = 3$ and $k = 3$, M_r $(r = 0, 1, 2)$ are H, T, and T', respectively, and each hole is a T. Therefore, we have the three-dimensional tiling by Hs and Ts. In the case $n = 4$ and $k = 2$, not only MCIs placed in lattice-cubes but also the holes are 16-cells, and we get the four-dimensional tiling by 16-cells. Since T and 16-cell are the only weak cross-polytope imaginary n-cube shapes for $n \geq 3$ (Proposition 6), among these tilings, there are only two tilings by imaginary cubes.

These two tilings are related to the fact that H and 16-cell are double imaginary cubes. The three-dimensional tiling by Hs and Ts can be characterized as follows [1]. Let σ^3 be the isometry on three-dimensional Euclidean space to rotate by 180 degrees around the axis $x = y = z$. Then, the tiling is a Voronoi tessellation of the union $\mathbb{Z}^3 \cup \sigma^3(\mathbb{Z}^3)$ of the two cubic lattices such that Voronoi cells of points in $\mathbb{Z}^3 \cap \sigma^3(\mathbb{Z}^3)$ have the shape H and those of other points have the shape T. See [6], for example, for the notion of Voronoi tessellations.

This construction can be extended to higher-dimensional cases. In the n-dimensional Euclidean space, let σ^n be the orthogonal transformation on $\mathbb{R}^n$ that satisfies $\sigma^n(1) = 1$ and $\sigma^n(v) = -v$ for $v \in \mathbb{R}^n$ with $v \cdot 1 = 0$. Then, take the Voronoi tessellation of $\mathbb{Z}^n \cup \sigma^n(\mathbb{Z}^n)$. The Voronoi cell of the origin is the intersection of two n-cubes $\mathrm{conv}(\{-1/2, 1/2\}^n)$ and $\sigma^n(\mathrm{conv}(\{-1/2, 1/2\}^n))$, and Voronoi cells of points in $\mathbb{Z}^n \cap \sigma^n(\mathbb{Z}^n)$ are its translations.

In the case $n = 4$, σ^4 maps the set V_1 to V_3, V_3 to V_1, and V_2 to itself, where the sets V_1, V_2, and V_3 are defined in Sect. 3.2. Therefore, the cube $\mathrm{conv}(\{-1/2, 1/2\}^4)) = \mathrm{conv}(V_2 \cup V_3)$ is mapped to the cube $\mathrm{conv}(V_2 \cup V_1)$ and their intersection $\mathrm{conv}(V_2)$ is the Voronoi cell at the origin. One can show that the other Voronoi cells are also 16-cells, and therefore this tiling is the four-dimensional tiling by 16-cells.

For $n \geq 3$, if the intersection E^n of two cubes $\mathrm{conv}(\{-1/2, 1/2\}^n)$ and $\sigma^n(\mathrm{conv}(\{-1/2, 1/2\}^n))$ is an imaginary cube of an n-cube C, then it must also be an imaginary cube of $\sigma^n(C)$. It is easy to show that C and $\sigma^n(C)$ are different n-cubes and thus E^n is a double imaginary cube. Since H and 16-cell are the only double imaginary n-cube shapes for $n \geq 3$ (Theorem 9), among these Voronoi tessellations there are only two tilings by imaginary cubes.

4   Fractal Imaginary Hypercubes

4.1   Fractal Imaginary Cubes

From a regular tetrahedron, one can form a fractal (i.e., self-similar) object known as a Sierpinski tetrahedron (Fig. 5(a)). It has similarity dimension two and it is also an imaginary cube.

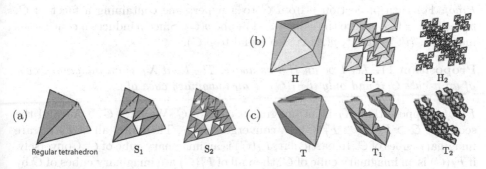

Fig. 5. The first two approximations of (a) Sierpinski tetrahedron, (b) H_∞, and (c) T_∞.

Let $\mathcal{H}^n$ be the metric space of non-empty compact subsets of $\mathbb{R}^n$ with the Hausdorff metric. According to the theory of IFS (iterated function system) fractals developed by Hutchinson [7], for contractions $f_i : \mathbb{R}^n \rightarrow \mathbb{R}^n$ $(i = 1, \ldots, m)$, an IFS $I = \{f_i \mid i = 1, 2, \ldots, m\}$ defines a fractal object as the fixedpoint of the following contraction map on $\mathcal{H}^n$:

$$F_I(X) = \bigcup_{i=1}^{m} f_i(X). \qquad (4)$$

As for a Sierpinski tetrahedron, let S be a regular tetrahedron and let $I_S = \{f_i : \mathbb{R}^3 \rightarrow \mathbb{R}^3 \mid i = 1, 2, 3, 4\}$ be an IFS where $f_i(i = 1, 2, 3, 4)$ are homothetic transformations (i.e., similitudes that perform no rotations) with the scale $1/2$ whose centers are vertices of S. The induced fractal is a Sierpinski tetrahedron. It is an imaginary cube of the cube C of which S is an imaginary cube. Note that this fractal object is minimal among imaginary cubes of C.

As generalizations of a Sierpinski tetrahedron, fractal imaginary cubes such that an IFS that induces the fractal is composed of k^2 homothetic transformations of scale $1/k$ are studied [2]. Sierpinski tetrahedron is the only such shape for $k = 2$. In the case $k = 3$, there are two such fractal shapes H_∞ and T_∞ whose convex hulls are H and T, respectively (Fig. 5(b,c)). In particular, H_∞ is a double imaginary cube. In the following, we explain these fractal imaginary cubes and their higher-dimensional counterparts.

For $k \geq 2$, let $I = \{f_i : \mathbb{R}^n \to \mathbb{R}^n \mid i = 1, 2, \dots, k^{n-1}\}$ be an IFS such that $f_i (i = 1, 2, \dots, k^{n-1})$ are homothetic transformations with the scale $1/k$. Let X_I be the fractal object obtained as the fixedpoint of the contraction map F_I on $\mathcal{H}^n$ defined by (4). Since X_I is the fixedpoint of F_I, for any $B \in \mathcal{H}^n$, the sequence $B, F_I(B), F_I^2(B), \dots$ converges to X_I with respect to the Hausdorff metric. Here, f^m is the m-times repetition of f.

Lemma 10. *Let C be an n-cube and let $(A_i; i = 0, 1, \dots)$ be a sequence of imaginary n-cubes of C. If the sequence $(A_i; i = 0, 1, \dots)$ converges to A with respect to the Hausdorff metric, then A is also an imaginary cube of C.*

Proof. For each projection p from C to a hyperplane containing a facet of C, $p(A_i)$ for $i = 0, 1, \dots$ are the same $(n-1)$-cube $p(C)$. Since p induces a continuous map from $\mathcal{H}^n$ to $\mathcal{H}^{n-1}$, $p(A)$ is also equal to $p(C)$.   □

Proposition 11. *Let I be an IFS as above. The limit X_I is an imaginary cube of an n-cube C if and only if $F_I(C)$ is an imaginary cube of C.*

Proof. Suppose that X_I is an imaginary cube of C. We have $C \supset X_I$ and the sequence $C \supset F_I(C) \supset F_I^2(C) \cdots$ converges to X_I. Therefore, all of $F_I^i(C)$ are imaginary cubes of C. In particular, $F_I(C)$ is an imaginary cube of C. Conversely, if $F_I(C)$ is an imaginary cube of C, then all of $F_I^i(C)$ are imaginary cubes of C by induction, and the limit X_I is also an imaginary cube of C from Lemma 10.   □

The fractal object X_I has the similarity dimension $n - 1$. Note that $F_I(C)$ is an imaginary cube of C if and only if $f_i(C)$ $(i = 1, 2, \dots, k^{n-1})$ are n-cubes obtained by cutting C into k^n n-cubes of the same size and selecting k^{n-1} of them so that they form an imaginary n-cube. Such a selection of k^{n-1} cubes corresponds to an $(n-1)$-dimensional Latin hypercube of order k. See, for example, [9] for the notion of a Latin hypercube.

4.2   Higher-Dimensional Extensions of the Sierpinski Tetrahedron

Let C be the n-cube $\mathrm{conv}(\{0,1\}^n)$. We set $P_a^n = \frac{1}{2}(C + \{a\})$ for $a \in \{0,1\}^n$. There are the following two ways of selecting 2^{n-1} n-cubes from $\{P_a^n \mid a \in \{0,1\}^n\}$ to form an imaginary cube.

$$\hat{S}^n = \cup\{P_a^n \mid a \in \{0,1\}^n, \, a \cdot 1 \equiv 0 \pmod 2\},$$
$$\hat{S}'^n = \cup\{P_a^n \mid a \in \{0,1\}^n, \, a \cdot 1 \equiv 1 \pmod 2\}.$$

Fig. 6. The three shapes $\hat{S}^3$, $\hat{H}^3$, and $\hat{T}^3$.

These two imaginary cubes have the same shape which we denote by $\hat{S}^n$ (Fig. 6). Let I_S be the IFS that consists of 2^{n-1} homothetic transformations with the scale $1/2$ that map C to the cubes in $\hat{S}^n$, and let S^n_∞ be the fractal induced by I_S. S^n_∞ is a fractal imaginary n-cube with the similarity dimension $n-1$ by Proposition 11. We denote by S^n_∞ the shape of S^n_∞. The shape S^3_∞ is the Sierpinski tetrahedron. Since all the components of I_S are homothetic transformations, the convex hull of S^n_∞ is equal to the convex hull of the centers of I_S, which is S^n defined in Sect. 3.3.

It is immediate to show that $\hat{S}^n$ and $\hat{S}'^n$ are the only two ways of selecting 2^{n-1} n-cubes from $\{P^n_a \mid a \in \{0,1\}^n\}$ to form an imaginary cube. Therefore, S^n_∞ is the only fractal imaginary cube shape obtained as the limit of an IFS that is composed of 2^{n-1} homothetic transformations with the scale $1/2$.

4.3 Fractal Imaginary Cubes H_∞ and T_∞ and Their Higher-Dimensional Extensions

We study the case $k = 3$. Let C be the n-cube $\mathrm{conv}(\{-1,1\}^n)$. We define $Q^n_a \subset C$ ($a \in \{-1,0,1\}^n$) as $\frac{1}{3}(C + \{2a\})$. There are the following three ways of selecting 3^{n-1} n-cubes from $\{Q^n_a \mid a \in \{-1,0,1\}^n\}$ to form an imaginary cube of C.

$$\hat{H}^n = \cup\{Q^n_a \mid a \in \{-1,0,1\}^n, \; a \cdot 1 \equiv 0 \pmod 3\},$$
$$\hat{T}^n = \cup\{Q^n_a \mid a \in \{-1,0,1\}^n, \; a \cdot 1 \equiv -1 \pmod 3\},$$
$$\hat{T}'^n = \cup\{Q^n_a \mid a \in \{-1,0,1\}^n, \; a \cdot 1 \equiv 1 \pmod 3\}.$$

$\hat{T}^n$ and $\hat{T}'^n$ have the same shape which we denote by $\hat{T}^n$. We denote by $\hat{H}^n$ the shape of $\hat{H}^n$ (Fig. 6).

Let I_H (resp. I_T) be the IFS that consists of 3^{n-1} homothetic transformations with the scale $1/3$ that map C to the cubes in $\hat{H}^n$ (resp. $\hat{T}^n$), and let H^n_∞ (resp. T^n_∞) be the fractal induced by I_H (resp. I_T). H^n_∞ and T^n_∞ are fractal imaginary n-cubes with the similarity dimension $n - 1$ by Proposition 11. We write H^n_∞ and T^n_∞ for their shapes.

The convex hull of H^n_∞ is equal to the convex hull of the centers of the components of I_H because they are homothetic transformations. It is the set

$$D^n = \{x \in \{-1,0,1\}^n \mid x \cdot 1 \equiv 0 \pmod 3\}.$$

From (3), the set of vertices of the polytope H^n defined in Sect. 3.3 is the intersection of D^n with the edges of C. Therefore, the convex hull of D^n coincides with H^n. Similarly, we can show that the convex hull of T^n_∞ is T^n.

Theorem 12. *For $n \geq 3$, H^n_∞ and T^n_∞ are the only fractal imaginary cube shapes obtained as the limit of an IFS that is composed of 3^{n-1} homothetic transformations with the scale $1/3$.*

Proof. Suppose that $n \geq 2$ and that $U \subset \{-1, 0, 1\}^n$ satisfies $\#U = 3^{n-1}$ and $\cup\{Q^n_a \mid a \in U\}$ is an imaginary cube of C. We show that there exist $b \in \{-1, 1\}^n$ and $r \in \{-1, 0, 1\}$ such that $U = U(b, r)$ for $U(b, r) = \{a \in \{-1, 0, 1\}^n \mid a \cdot b \equiv r \pmod 3\}$. It is clear that such a selection U is congruous to that of $\hat{H}^n$ or $\hat{T}^n$. We show this by induction on n, and it is true for $n = 2$.

Note that we have $U(b, r) = U(b', r')$ if and only if $(b, r) = \pm(b', r')$. Since simultaneous equations $a_1 + a_2 \equiv r_1, a_1 - a_2 \equiv r_2 \pmod 3$ always have a solution $(a_1, a_2) = 2(r_1 + r_2, r_1 - r_2)$, one can also find that if $b \neq \pm b'$, then we have $U(b, r) \cap U(b', r') \neq \emptyset$ for any choice of $r, r' \in \{-1, 0, 1\}$.

Suppose that $n \geq 3$. We divide U into three parts

$$U = U_{-1} \times \{-1\} \cup U_0 \times \{0\} \cup U_1 \times \{1\},$$

where $U_i \subset \{-1, 0, 1\}^{n-1}$ satisfies $\#U_i = 3^{n-2}$ and that $\cup\{Q^{n-1}_a \mid a \in U_i\}$ is an imaginary $(n-1)$-cube for $i \in \{-1, 0, 1\}$. From the assumption, we can put $U_i = U(b_i, r_i)$ for $i \in \{-1, 0, 1\}$. Considering the projection in the n-th direction, we have $U_i \cap U_j = \emptyset$ for $-1 \leq i < j \leq 1$. Therefore, we can assume that $b_{-1} = b_0 = b_1 = (b_1, \ldots, b_{n-1})$, and we get $\{r_{-1}, r_0, r_1\} = \{-1, 0, 1\}$. In each case, there is $b_n \in \{-1, 1\}$ such that $r_0 \equiv r_{-1} - b_n \equiv r_1 + b_n \pmod 3$, and hence we obtain $U = U((b_1, \ldots, b_n), r_0)$.   □

References

1. Tsuiki, H.: Imaginary cubes and their puzzles. Algorithms **5**, 273–288 (2012)
2. Tsuiki, H.: Does it look square – hexagonal bipyramids, triangular antiprismoids, and their fractals. In: Proceedings of Conference Bridges Donostia, pp. 277–286. Tarquin Publications (2007)
3. Tsuiki, H.: Imaginary cubes–objects with three square projection images. In: Proceedings of Bridges 2010: Mathematics, Music, Art, Architecture, Culture, pp. 159–166. Tessellations Publishing, Phoenix (2010)
4. Tsuiki, H.: SUDOKU colorings of the hexagonal bipyramid fractal. In: Ito, H., Kano, M., Katoh, N., Uno, Y. (eds.) KyotoCGGT 2007. LNCS, vol. 4535, pp. 224–235. Springer, Heidelberg (2008)
5. Ziegler, G.M.: Lectures on Polytopes. Springer, New York (1995)
6. Okabe, A., Boots, B., Sugihara, K.: Spatial Tessellations Concepts and Applications of Voronoi Diagrams. Wiley, New York (1992)
7. Hutchinson, J.: Fractals and self similarity. Indiana Univ. Math. J. **30**, 713–747 (1981)
8. Coxeter, H.S.M.: Regular Polytopes. Dover, New York (1973)
9. McKay, B.D., Wanless, I.M.: A census of small latin hypercubes. SIAM J. Discrete Math. **22**, 719–736 (2008)

More Results on Clique-chromatic Numbers
of Graphs with No Long Path

Tanawat Wichianpaisarn and Chariya Uiyyasathian[✉]

Faculty of Science, Department of Mathematics and Computer Science,
Chulalongkorn University, Payathai Rd., Bangkok 10330, Thailand
tanawat.wp@gmail.com, chariya.u@chula.ac.th

Abstract. The clique-chromatic number of a graph is the least number of colors on the vertices of the graph so that no maximal clique of size at least two is monochromatic. In 2003, Gravier, Hoang, and Maffray have shown that, for any graph F, the class of F-free graphs has a bounded clique-chromatic number if and only if F is a vertex-disjoint union of paths, and they give an upper bound for all such cases. In this paper, their bounds for $F = P_2 + kP_1$ and $F = P_3 + kP_1$ with $k \geq 3$ are significantly reduced to $k + 1$ and $k + 2$ respectively, and sharp bounds are given for some subclasses.

Keywords: Clique-chromatic number · Clique-coloring

2010 Mathematics Subject Classification: 05C15

1 Introduction

All graphs considered in this paper are simple. We use terminologies from West's textbook [9]. $V(G)$ and $E(G)$ denote the vertex set and the edge set of a graph G, respectively. The symbols K_n, P_n and C_n denote the complete graph, path, and cycle, with n vertices, respectively. The *diamond* is the complete graph K_4 minus an edge. The *neighborhood* of a vertex x in a graph G is the set of vertices adjacent to x, and is denoted by $N_G(x)$. For $S \subseteq V(G)$, $N_S(x)$ stands for the neighborhood of a vertex x in S, that is, $N_S(x) = N_G(x) \cap S$. Given graphs $G_1, G_2, \ldots, G_k$ with pairwise disjoint vertex sets, the *disjoint union* of graphs $G_1, G_2, \ldots, G_k$ is the graph with vertex set $\bigcup_{i=1}^{k} V(G_i)$ and edge set $\bigcup_{i=1}^{k} E(G_i)$, denoted by $G_1 + G_2 + \cdots + G_k$. For $k \in \mathbb{N}$, kG is the disjoint union of k pairwise disjoint copies of a graph G.

A subset Q of $V(G)$ is a *clique* of G if any two vertices of Q are adjacent. A clique is *maximal* if it is not properly contained in another clique. A *k-coloring* of a graph G is a function $f : V(G) \rightarrow \{1, 2, \ldots, k\}$. A *proper k-coloring* of a graph G is a k-coloring of G such that adjacent vertices have different colors. The *chromatic number* of a graph G is the smallest positive integer k such that

Tanawat Wichianpaisarn—Partially supported by His Royal Highness Crown Prince Maha Vajiralongkorn Fund.

J. Akiyama et al. (Eds.): JCDCGG 2013, LNCS 8845, pp. 185–190, 2014.
DOI: 10.1007/978-3-319-13287-7_16

G has a proper k-coloring, denoted by $\chi(G)$. A *proper k-clique-coloring* of a graph G is a k-coloring of G such that no maximal clique of G with size at least two is monochromatic. A graph G is *k-clique-colorable* if G has a proper k-clique-coloring. The *clique-chromatic number* of G is the smallest k such that G has a proper k-clique-coloring, denoted by $\chi_c(G)$. Note that $\chi_c(G) = 1$ if and only if G is an edgeless graph. Since any proper k-coloring of G is a proper k-clique-coloring of G, $\chi_c(G) \leq \chi(G)$. Recall that a *triangle* is the complete graph K_3. If G is a triangle-free graph, then maximal cliques of G are edges; so $\chi_c(G) = \chi(G)$. Mycielski [8] showed that the family of triangle-free graphs has no bounded chromatic number. Consequently, it has no bounded clique-chromatic number, either. On the other hand, many families of graphs have bounded clique-chromatic numbers, for example, comparability graphs, cocomparability graphs, and the k-power of cycles (see [2,4,5]). In 2004, Bacso et al. [1] proved that almost all perfect graphs are 3-clique-colorable and conjectured that all perfect graphs are 3-clique-colorable.

A subgraph H of a graph G is said to be *induced* if, for any pair of vertices x and y of H, xy is an edge of H if and only if xy is an edge of G. For a given graph F, a graph G is *F-free* if it does not contain F as an induced subgraph. A graph G is $(F_1, F_2, \ldots, F_k)$-*free* if it is F_i-free for all $1 \leq i \leq k$. In [6], Gravier, Hoang and Maffray gave a significant result that, for any graph F, the family of all F-free graphs has a bounded clique-chromatic number if and only if F is a vertex-disjoint union of paths. Many authors explored more results in $(F_1, F_2, \ldots, F_k)$-free graphs. Gravier and Skrekovski [7] in 2003 proved that $(P_3 + P_1)$-free graphs unless it is C_5, and (P_5, C_5)-free graphs are 2-clique-colorable. In 2004, Bacso et al. [1] showed that (claw, odd hole)-free graphs are 2-clique-colorable. Later, Defossez [3] in 2006 proved that (diamond, odd hole)-free graphs are 4-clique-colorable, and (bull, odd hole)-free graphs are 2-clique-colorable.

Given a graph F, let $f(F) = \max\{\chi_c(G) \mid G \text{ is an } F\text{-free graph}\}$. When F_1 is an induced subgraph of F_2, if a graph G is F_1-free then G is also F_2-free, it follows that $f(F_1) \leq f(F_2)$. In 2003, Gravier, Hoang and Maffray [6] showed the following result.

Theorem 1 [6]. *Let F be a graph. Then $f(F)$ exists if and only if F is a vertex-disjoint union of paths. Moreover,*

– *if $|V(F)| \leq 2$ or $F = P_3$ then $f(F) \leq 2$,*
– *else $f(F) \leq cc(F) + |V(F)| - 3$ where $cc(F)$ is the number of connected components of a graph F.*

Furthermore, they proved that $(P_2 + 2P_1)$-free graphs and $(P_3 + 2P_1)$-free graphs are 3-clique-colorable. Since the cycle C_5 is both $(P_2 + 2P_1)$-free and $(P_3 + 2P_1)$-free with $\chi_c(C_5) = 3$, this bound is sharp.

2   Main Results

An *independent set* in a graph is a set of pairwise nonadjacent vertices. A *maximum independent set* of a graph G is a largest independent set of G and its size is denoted by $\alpha(G)$. Bacso et al. [1] stated the relationship between the clique-chromatic number and the size of a maximum independent set of a graph, as follows.

Theorem 2 [1]. *Let G be a graph. If $G \neq C_5$ and G is not a complete graph, then $\chi_c(G) \leq \alpha(G)$.*

It follows from Theorem 1 that every $(P_2 + kP_1)$-free graph is $(2k)$-clique-colorable and every (P_3+kP_1)-free graph is $(2k+1)$-clique-colorable. We improve these upper bounds for $k \geq 3$.

Theorem 3. *For $k \geq 3$, a $(P_2 + kP_1)$-free graph is $(k + 1)$-clique-colorable.*

Proof. Let G be a $(P_2 + kP_1)$-free graph. Let $S = \{s_0, s_1, \ldots, s_{\alpha(G)-1}\}$ be a maximum independent set of G. If $\alpha(G) \leq k$, then $\chi_c(G) \leq k$ by Theorem 2.

Assume $\alpha(G) \geq k + 1$. Let $M(s_0) = V(G) \backslash (S \cup N_G(s_0))$. For $R \subseteq S \backslash \{s_0\}$, define $Y_R = \{v \in M(s_0) \mid N_S(v) = S \backslash (\{s_0\} \cup R)\}$ and $\min(R) = \min\{i \in \mathbb{N} \mid s_i \notin R\}$. In particular, $\min(\emptyset) = 1$. Note that $V(G)$ is the disjoint union of S, $N_G(s_0)$ and Y_R where $R \subseteq S \backslash \{s_0\}$. Let f be the coloring of G defined by

$$f(v) = \begin{cases} 1, & \text{if } v \in S \\ 2, & \text{if } v \in N_G(s_0) \\ \min(R) + 2, & \text{if } v \in Y_R \text{ where } R = S \backslash (\{s_0\} \cup N_S(v)). \end{cases}$$

Now, let $R \subseteq S \backslash \{s_0\}$ where $Y_R \neq \emptyset$, and let $y \in Y_R$. If $R = S \backslash \{s_0\}$, then $N_S(y) = \emptyset$; so $S \cup \{y\}$ is an independent set of G. This contradicts the maximality of S. Thus $R \neq S \backslash \{s_0\}$. If $|R| \geq k-1$, then the subgraph of G induced by $S \cup \{y\}$ contains an induced subgraph $P_2 + kP_1$, a contradiction. Thus $|R| \leq k - 2$, and it follows that $\min(R) \leq k - 1$. Therefore, f is a $(k + 1)$-coloring of G. Suppose that G has a monocolored maximal clique Q of size at least two, say colored by m. Since S is an independent set, $m \neq 1$. Thus $Q \cap S = \emptyset$. Note that $s_{\min(R)}$ is adjacent to all vertices of Y_R. Thus s_{m-2} is adjacent to all vertices of Q. Then $Q \cup \{s_{m-2}\}$ is a clique of G. It contradicts the maximality of Q. Hence $\chi_c(G) \leq k + 1$.

Theorem 4. *For $k \geq 3$, a $(P_3 + kP_1)$-free graph is $(k + 2)$-clique-colorable.*

Proof. Let G be a $(P_3 + kP_1)$-free graph. Let $S = \{s_1, s_2, \ldots, s_{\alpha(G)}\}$ be a maximum independent set of G. If $\alpha(G) \leq k + 1$, then $\chi_c(G) \leq k + 1$ by Theorem 2. Assume $\alpha(G) \geq k+2$. For $1 \leq i \leq \alpha(G)$, let $X_i = \{v \in V(G) \backslash S \mid N_S(v) = \{s_i\}\}$. Suppose that there is an edge, say $x_i x_j$, between X_i and X_j where $i \neq j$. Then there exist k vertices in $S \backslash \{s_i, s_j\}$ together with s_i, x_i, x_j form an induced subgraph P_3+kP_1 of G, a contradiction. Thus there is no edge between any two X_i's. For $R \subseteq S$ where $|R| \neq \alpha(G) - 1$, define $Y_R = \{v \in V(G) \backslash S \mid N_S(v) = S \backslash R\}$ and $\min(R) = \min\{i \in \mathbb{N} \mid s_i \notin R\}$. Note that $V(G)$ is the disjoint union of

S, X_i where $1 \leq i \leq \alpha(G)$, and Y_R where $R \subseteq S$ and $|R| \neq \alpha(G) - 1$. Let f be the coloring of G defined by

$$f(v) = \begin{cases} 1, & \text{if } v \in S \\ 2, & \text{if } v \in \bigcup_{i=1}^{\alpha(G)} X_i \\ \min(R) + 2, & \text{if } v \in Y_R \text{ where } R = S \backslash N_S(v). \end{cases}$$

Let $R \subseteq S$ where $Y_R \neq \emptyset$, and let $y \in Y_R$. If $R = S$, then $N_S(y) = \emptyset$; so $S \cup \{y\}$ is an independent set of G, a contradiction. If $k \leq |R| \leq \alpha(G) - 2$, then the subgraph of G induced by $S \cup \{y\}$ contains an induced subgraph $P_3 + kP_1$, a contradiction. Thus $|R| \leq k - 1$, and it follows that $\min(R) \leq k$. Hence f is a $(k+2)$-coloring of G. Now, suppose that G has a monocolored maximal clique Q of size at least two, say colored by m. Since S is an independent set, $m \neq 1$. If $m = 2$, then $Q \subseteq X_i$ for some i. We have that s_i is adjacent to all vertices of Q, a contradiction. Now, assume $m \geq 3$. Since $s_{\min(R)}$ is adjacent to all vertices of Y_R, s_{m-2} is adjacent to all vertices of Q, a contradiction. Thus f is a proper $(k+2)$-clique-coloring of G, and hence $\chi_c(G) \leq k + 2$.

Theorem 3 ensures that every $(P_2 + kP_1)$-free graph where $k \geq 3$ is $(k+1)$-clique-colorable but we have found no graph guaranteeing this sharpness yet. However, when $k = 3$ and 4, there is a $(P_2 + kP_1)$-free graph which is k-clique-colorable, namely, the cycle C_5 is $(P_2 + 3P_1)$-free and $\chi_c(C_5) = 3$, and the 4-chromatic Mycielski's graph G_4 [8] is $(P_2 + 4P_1)$-free and $\chi_c(G_4) = 4$. (See Fig. 1) Notice that both of them are diamond-free, this suggests the result in Theorem 5.

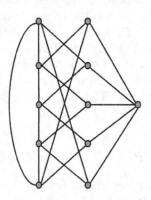

Fig. 1. The 4-chromatic Mycielski's graph G_4

Theorem 5. *For $k \geq 3$, a $(P_2 + kP_1$, diamond)-free graph is k-clique-colorable.*

Proof. Let G be a $(P_2 + kP_1$, diamond)-free graph. If $\alpha(G) \leq k$, then $\chi_c(G) \leq k$ by Theorem 2. Assume $\alpha(G) \geq k+1$. Use the same terminologies and arguments as in the proof of Theorem 3, we can define a k-coloring of G as follows:

$$g(v) = \begin{cases} 1, & \text{if } v \in S \\ 2, & \text{if } v \in N_G(s_0) \\ \min(R) + 2, & \text{if } v \in Y_R \text{ where } R = S\backslash(\{s_0\} \sqcup N_{\mathcal{G}}(v)) \text{ and} \\ & \min(R) \le k - 2 \\ k, & \text{if } v \in Y_R \text{ where } R = S\backslash(\{s_0\} \cup N_S(v)) \text{ and} \\ & \min(R) = k - 1. \end{cases}$$

To claim that g is a proper k-clique-coloring of G, suppose that G has a monocolored maximal clique Q of size at least two, say colored by m. Since S is an independent set, $m \ne 1$. If $m \le k - 1$, then s_{m-2} is adjacent to all vertices of Q, a contradiction. Assume $m = k$. Then $Q \subseteq \bigcup\{Y_R \mid R \subseteq S\backslash\{s_0\}$ and $k - 2 \le \min(R) \le k - 1\}$. Since $Y_R = \emptyset$ for all $R \subseteq S\backslash\{s_0\}$ where $|R| \ge k - 1$, we consider only Y_R where $|R| \le k - 2$. Thus if $k - 2 \le \min(R) \le k - 1$, then $R = \{s_1, s_2, \ldots, s_{k-3}, s_t\}$ where $k - 2 \le t \le \alpha(G) - 1$. Since G is diamond-free and $\alpha(G) - 1 \ge k$, Y_R is an independent set, and then $|Q \cap Y_R| \le 1$ for each $R \subseteq S\backslash\{s_0\}$. If $|Q| \ge 3$, then there exists a diamond induced by a vertex in $S\backslash\{s_0\}$ and three vertices in Q, a contradiction. So $|Q| = 2$. Let $Q \subseteq Y_{R_1} \cup Y_{R_2}$ for some $R_1, R_2 \subseteq S\backslash\{s_0\}$ where $R_1 \ne R_2$ and $k - 2 \le \min(R_1), \min(R_2) \le k - 1$. Then $|R_1 \cup R_2| \le k - 1$. Since $\alpha(G) - 1 \ge k$, there exists a vertex in $S\backslash\{s_0\}$ that is adjacent to both vertices of Q, a contradiction. Hence $\chi_c(G) \le k$.

Similarly to $(P_2 + kP_1)$-free graphs, the result for $(P_3 + kP_1)$-free graphs in Theorem 4 has not been proved to be sharp. Theorem 6 gives its subclass of graphs using at most $k + 1$ colors.

Theorem 6. *For $k \ge 3$, a $(P_3 + kP_1, diamond)$-free graph is $(k + 1)$-clique-colorable.*

Proof. Let G be a $(P_3 + kP_1, diamond)$-free graph. If $\alpha(G) \le k+1$, then $\chi_c(G) \le k + 1$ by Theorem 2. Assume $\alpha(G) \ge k + 2$. Use the same terminologies and arguments as in the proof of Theorem 4, we can define a $(k + 1)$-coloring of G as follows:

$$g(v) = \begin{cases} 1, & \text{if } v \in S \\ 2, & \text{if } v \in \bigcup_{i=1}^{\alpha(G)} X_i \\ \min(R) + 2, & \text{if } v \in Y_R \text{ where } R = S\backslash N_S(v) \text{ and } \min(R) \le k - 1 \\ k + 1, & \text{if } v \in Y_R \text{ where } R = S\backslash N_S(v) \text{ and } \min(R) = k. \end{cases}$$

Suppose that G has a monocolored maximal clique Q of size at least two, say colored by m. If $m = 2$, then $Q \subseteq X_i$ for some i; so s_i is adjacent to all vertices of Q, a contradiction. If $3 \le m \le k$, then s_{m-2} is adjacent to all vertices of Q, a contradiction. Assume $m = k + 1$. Then $Q \subseteq \bigcup\{Y_R \mid R \subseteq S$ and $k - 1 \le \min(R) \le k\}$. Since G is diamond-free and $\alpha(G) \ge k + 2$, Y_R is an independent set. Thus $|Q \cap Y_R| \le 1$ for each $R \subseteq S$. If $|Q| \ge 3$, then there exist a vertex in S together with any three vertices in Q which induce a diamond, a contradiction. So $|Q| = 2$. Since $\alpha(G) \ge k + 2$, there exists a vertex in S that is adjacent to both vertices of Q, a contradiction. Hence $\chi_c(G) \le k + 1$.

Since the 4-chromatic Mycielski's graph G_4 is $(P_3 + 3P_1, \text{diamond})$-free, the upper bound in Theorem 6 for the case $k = 3$ is sharp.

References

1. Bacsó, G., Gravier, S., Gyárfás, A., Preissmann, M., Sebő, A.: Coloring the maximal cliques of graphs. SIAM J. Discrete Math. **17**, 361–376 (2004)
2. Campos, C.N., Dantasa, S., de Mello, C.P.: Colouring clique-hypergraphs of circulant graphs. Electron. Notes Discrete Math. **30**, 189–194 (2008)
3. Defossez, D.: Clique-coloring some classes of odd-hole-free. J. Graph Theory **53**, 233–249 (2006)
4. Duffus, D., Sands, B., Sauer, N., Woodrow, R.E.: Two-coloring all two-element maximal antichains. J. Comb. Theory Ser. A **57**, 109–116 (1991)
5. Duffus, D., Kierstead, H.A., Trotter, W.T.: Fibres and ordered set coloring. J. Comb. Theory Ser. A **58**, 158–164 (1991)
6. Gravier, S., Hoáng, C.T., Maffray, F.: Coloring the hypergraph of maximal cliques of a graph with no long path. Discrete Math. **272**, 285–290 (2003)
7. Gravier, S., Škrekovski, R.: Coloring the clique hypergraph of graphs without forbidden structure, Les cahiers du laboratoire Leibniz **83**, (2003). http://www-leibniz. imag.fr/LesCahiers/
8. Mycielski, J.: Sur le coloriage des graphes. J. Colloq. Math. **3**, 161–162 (1955)
9. West, D.B.: Introduction to Graph Theory. Prentice Hall, New Jersey (2001)

Author Index

Printed in the United States
By Bookmasters